The MINES of NEWENT and ROSS

David Bick

Published by THE POUND HOUSE, Newent, Glos

The Mines of Newent and Ross

By the same author
The Gloucester & Cheltenham Railway 1968 (Enlarged edition 1987)
**Old Leckhampton* 1971
**Dylife*1975 (New edition 1985)
**The Old Metal Mines of Mid-Wales* (1974-8)
**The Hereford & Gloucester Canal* 1979
**The Old Industries of Dean* 1980
**The Old Copper Mines of Snowdonia* 1982 (New edition 1985)
Frongoch Lead and Zinc Mine 1986
Sygun Copper Mine 1987
The Gold Mines of Wales - in preparation

The author wishes to emphasise that some sites are potentially dangerous, and that permission should be obtained before venturing on private property.

ISBN 0 906885 06 X

Cover picture: Newent Colliery in 1879 - a reconstruction by Michael Blackmore.

*Published by THE POUND HOUSE

Preface

The mines of Newent and district, of iron and coal and even trials for silver and gold, were active long before the Industrial Revolution and did not finally cease until the present century. For generations they furnished the local ironworks with ore, and in the 1790s the little Newent Coalfield was deemed of sufficient promise to attract a canal to the town.

All this long chapter is nowadays forgotten, but for me, with remnants of old workings still to be discerned in hedgerow and thicket, it is a period of endless mystery and fascination.

When I began to investigate over thirty years ago, according to the best authority no records worthy of the name existed; an assertion which has proved pretty well correct. Nonetheless by dint of perseverance the semblance of a story began to emerge, and now, with a greatly awakening public interest in our heritage, I have felt it almost a duty to publish this small edition notwithstanding its abundant shortcomings. I hope it will fill a gap in the chronicles of Gloucestershire's industrial past, and form a memorial to those who followed a falling star.

I have also taken the opportunity to include a brief account of the mine workings in Penyard Park, near Ross, an intriguing subject of which very little is known.

Finally, a word of explanation. To be consistent with the times described, Imperial units have been used throughout and where necessary, metric measurements converted to them.

David Bick
Newent, 1987

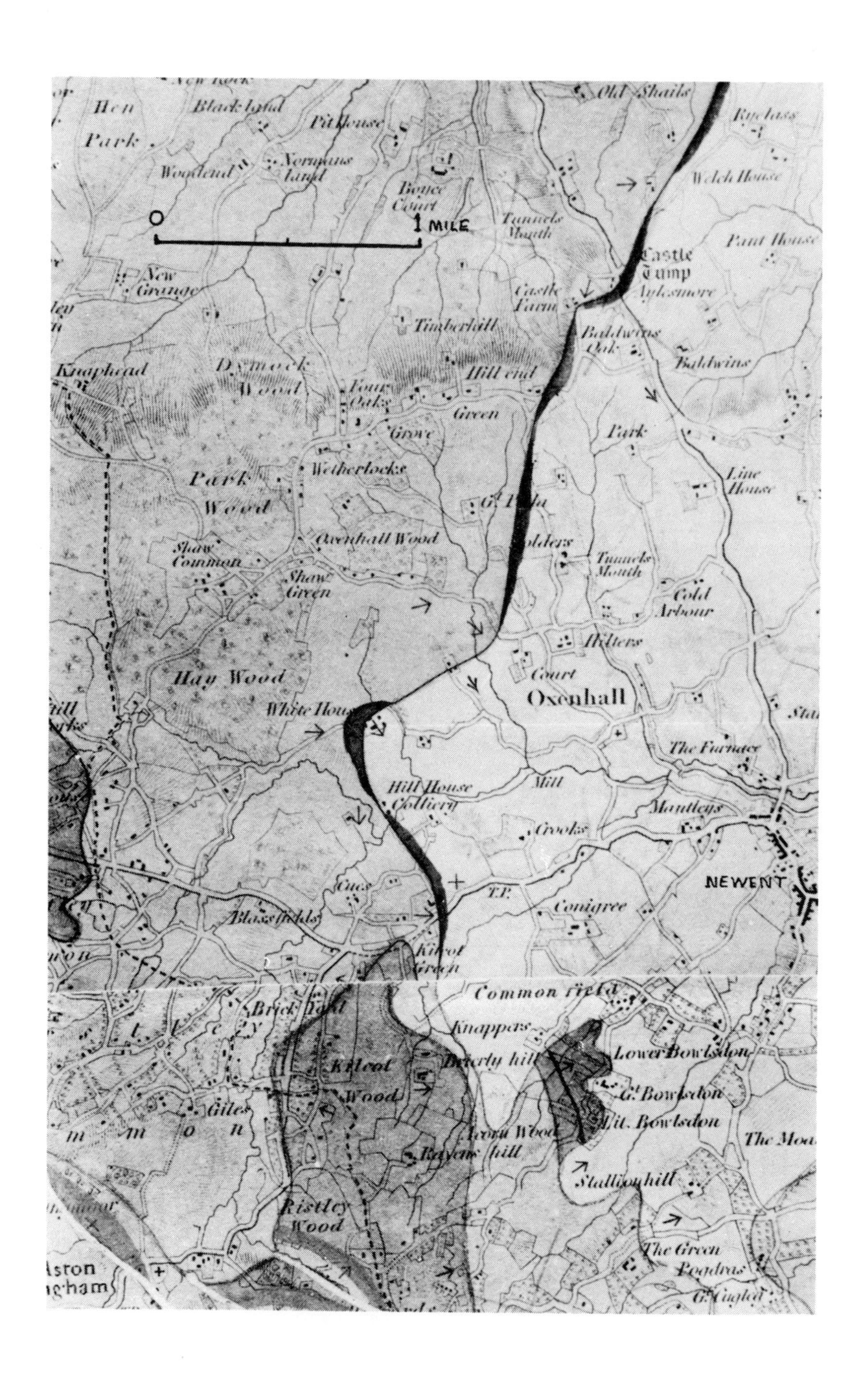

Hen
Park
Pithouse
Normans land
Woodend
Boyce Court
Old Shails
Welch House
O
1 MILE
Tunnels Mouth
New Grange
Castle Tump
Aylesmore
Castle Farm
Pant House
Timberhill
Baldwins Oak
Baldwins
Knaphead
Dymock Wood
Four Oaks
Hill end
Green
Grove
Park
Line House
Park Wood
Oxenhall Wood
Shaw Common
Shaw Green
Tunnels Mouth
Cold Arbour
Hilters
Hay Wood
Court
Oxenhall
White Hou
The Furnace
Hill House Colliery
Mill
Mantleys
Crooks
NEWENT
T.P.
Conigree
Blazefields
Kilcot Green
Common Field
Knappers
Lower Bowlsdon
Kilcot Wood
Gt. Bowlsdon
Lit. Bowlsdon
Giles
Acorn Wood
Ravens hill
Stallionhill
Ristley Wood
The Green

Contents

Opposite. A detail from the Geological Survey map of 1845, based on the Ordnance map of 1831. A new edition is imminently expected. Note the sinuous line of Coal Measures separating the Triassic (New Red) sandstones on the east from the Old Red formations on the west. To the south are Silurian rocks near May Hill.

Introduction

Although the mines of Newent and district were small, even insignificant compared to the neighbouring Forest of Dean, they were for centuries part of local industry. Their antiquity is not in doubt, for coal and iron can still be found just below the turf, and must have been early discovered. Ancient smelting operations known as bloomeries existed, and some may have predated Christianity.

The iron ores occurred in quite different geological conditions to those of Dean. They were worked in hematized Triassic sandstones overlaying the Coal Measures, and in older Silurian rocks to the south, where the geology is very faulted and complex. This forgotten source lies near May Hill and was worked for generations, well before the Industrial Revolution.

Together with cinder-tips which contained much residual metal from the old bloomeries, these ores contributed to the establishment of Elmbridge blast-furnace at Newent, in or before 1639. It grew into an important ironworks and did not close till 1751. Strangely enough, the deposits near May Hill have been ignored by almost every writer, and their mineralogy is entirely a mystery. Such workings are an instance of outcrops long since exploited and forgotten and leaving few, if any, traces behind. Where for example, was the 'minery in the baylywick of Leye [Lea]', which Richard Talebot claimed in 1657, or the deposit of iron ore on Gorsley Common?[1]

Although it is a tradition that the sandstone ores of Oxenhall helped to feed Elmbridge Furnace, there is little evidence in support. They were in general too siliceous, but rich goethite/hematite which occurs as far east as Pool Hill no doubt received attention in bloomery days. In the 19th century attemps to work these mainly low-grade ores met with little success, and it may be said that the iron mines and quarries of Newent and district died with the smelteries they had served for so long.

Traces of the workings may still be found, and as for blast-furnace slag, it was much used for hard-core and in gateways. There is scarcely a field within several miles of Newent where its vitreous grey/green and blue presence is not revealed by the plough.

Since the ironworks used charcoal fuel, Newent coal was of no account for that purpose; it was not until the vision of the Hereford & Gloucester Canal appeared with the prospect of cheap transport, that attention really focused on the possibilities. Whether in economic terms the coal raised exceeded in value that of the iron ore, we can be sure that in terms of profit it did not, for whereas the charcoal iron industry brought wealth to its investors, the collieries were almost without exception a failure.

It would be proper to devote a proportional space to both forms of mining, but the dearth of records relating to that of iron renders such a thing impossible. The major part of this volume will therefore be devoted to coal, though even here the records are also incomplete.

The Coal Measures, or strata in which seams occur, are fairly accurately represented on the Old Series one-inch geological map which is reproduced here, and still, at the time of writing, the most up-to-date available on such a scale. The main error is in the under-estimation south of Boulsdon, where carboniferous rocks extend (though barren of coal) almost to Green Farm and Great Cugley. Between Boulsdon and Knapper's Farm runs a major fault and to the north the width of exposures is much less. They follow a sinuous course via Kilcot, White House, Peter's Farm, Hillend and Castle Tump, finally fading out east of Dymock. The whole extent is four or five miles through the parishes of Newent, Oxenhall and Dymock, and included before the introduction of boundary changes, an outlying portion of Pauntley, near Kilcot.

A number of coal seams or veins were often encountered in a single shaft, and thin coals with partings of shale all close together were often classified as one thick seam, not least to give a good impression.

Newent coal tended to a high ash content and was rather friable and brittle. This vulnerability to rough handling did little to aid its chances, though its sulphurous nature has perhaps been exaggerated. Such an impurity did not seriously impede Parkend (Forest of Dean) coal which sold as far afield as Cornwall and Ireland, notwithstanding that 'it contained iron pyrites in very large quantities in every joint, even when broken into the smallest pieces.' *

The Coal Measures are sandwiched between the Old Red Sandstone formations consisting of marls and siltstones on the west and Triassic or Bromsgrove (New Red) Sandstone on the east. The latter imparts a characteristic colour and form to the landscape and can be readily observed in ancient sunken lanes as at Lower Boulsdon, White House and Holder's Lane. At their base and just above the Coal Measures these beds turn into a coarse conglomerate or pudding-stone that is sometimes of a yellowish colour. It is exposed near Knapper's Farm and in a quarry near Peter's Farm, also at Castle Tump and behind Merehills in Welsh House Lane. The sandstone is also frequently heavily charged with iron, as though descending ferriferous solutions were arrested by impervious clays beneath. Pockets of pure geothite (hydrated iron-oxide) also occurred, often displaying a beautiful fibrous or botryoidal structure. These deposits were rich enough in places to justify mining.

The Triassic sandstones dip easterly and usually only slightly, though sometimes reaching 30°. The Old Red dips steeper, even approaching vertical; in general the Coal Measures rest on them and conform to their inclination. This junction of rocks of widely differing age is associated with a north-south fault extending many miles and passing just east of the Malverns. It affects the whole of the Newent area and virtually every mining venture ran into trouble due to extensive subsidiary fractures.

It was long ago suspected that these shattered seams heralded vast deposits beneath the Trias of the Severn Vale, an important question which has been debated for nearly two centuries. In 1822 experienced geologists[2] could only report that the coalfield is 'concealed by overlying strata of the newer red sandstone and its relations have not as yet been distinctly ascertained' - a situation which in some respects still applies today. One of the first to consider

* See footnote, page 83

the matter was Roderick Murchison who began life as a fox-hunting aristocrat and fought with Wellington in Spain before becoming absorbed by the infant science of geology. Staying at the George Hotel he examined the district in 1833/4 and his researches will be much quoted. As long ago as the 1860s the British Government set up a Commission to study the subject of national coal reserves and in its report another very capable geologist, John Phillips, expressed a caution as to whether any coal extended to any marked degree east of the town.[3] In recent years deep borings[4] and gravity surveys have largely confirmed his doubts.

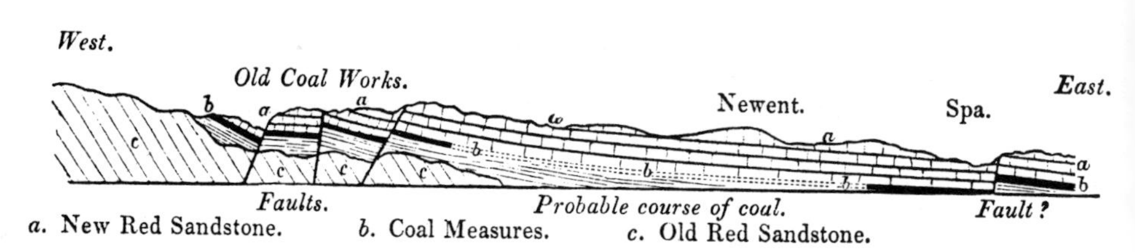

A section through the coalfield, from Murchison's *Silurian System*, 1839.

It is now believed that the measures which once existed in a continuous stratum from the Forest of Dean through Newent to the Forest of Wyre and eastwards under and beyond the Vale of Severn were subsequently broken up by late-Carboniferous movements and largely removed by erosion. In later ages Triassic deposits over a mile thick covered them so that development in future times is unlikely, even supposing substantial areas remain intact. Near Burford however, the beds are more complete and fairly shallow, forming part of the Oxford Coalfield as yet unworked. Attempts to correlate Newent coal with seams elsewhere were made as early as 1812 by the Forest of Dean scientist and ironmaster David Mushet.[6] After visiting Boulsdon he confidently stated that its seams equated with the Derbyshire thick first coal, and the 'small coals found under the Greasly Sheanick and Alfreton Coals. This curious circumstance was further corroborated by finding fragments of well-known argillaceous ironstone amongst the spoil.' Surprisingly enough, it now appears that a Midlands connection may not be beyond the bounds of possibility, at least for the areas north of Boulsdon. The Stallion Hill - Boulsdon area is more allied to the Forest of Dean and Mamble (Pensax) coalfields, and the overall age is Upper Coal Measures.[7] For its size, the Newent Coalfield is probably the most complex in England.

Turning now to the coalfields development, the extremely bad state of the roads proved a major obstacle. In 1779 the historian Samuel Rudder was obliged 'to desert his horse and to travel with a guide from one village to another.' Thus the promotion in 1790 of the Hereford & Gloucester Canal with a branch to Newent[8] was strongly welcomed by landowners including John Nourse Morse and the Hon. Andrew Foley who lived at Ameley in Herefordshire but owned practically the whole of Oxenhall.

A canal between the two cities had long been desired and much was made of potential coal traffic. According to the promotors, 'It is generally acknowledged that the City of Hereford and its environs are great sufferers from the want of a regular supply of coal. That there are considerable quanitities of this commodity of a superior quality in the neighbourhood of Newent is an unquestionable matter of fact . . .'[9]

With considerable publicity a wagon-load of free coal went to Gloucester,[10] and early in 1791 those attending a meeting in Hereford learnt that additional veins of coal had been discovered 'so as to leave no apprehension of a scarcity.'

At this period canals as an investment were highly regarded, and the company gained its Act in April 1791. Prospects appeared so promising that the directors decided to make the best of the coalfield by re-aligning the route close to Newent itself instead of following the much easier valley of the Leadon. In fact, both interests were increasingly seeking salvation in a touching display of mutual faith, and for a time it seemed as if the dream might be fulfilled. The canal company

A typical small colliery *Antique Collector's Club*
The Newent pits were no doubt similar to this illustration by J. C. Ibbetson of a Forest of Dean colliery about 1800. Note the horse-whim and fumes issuing from the shaft.

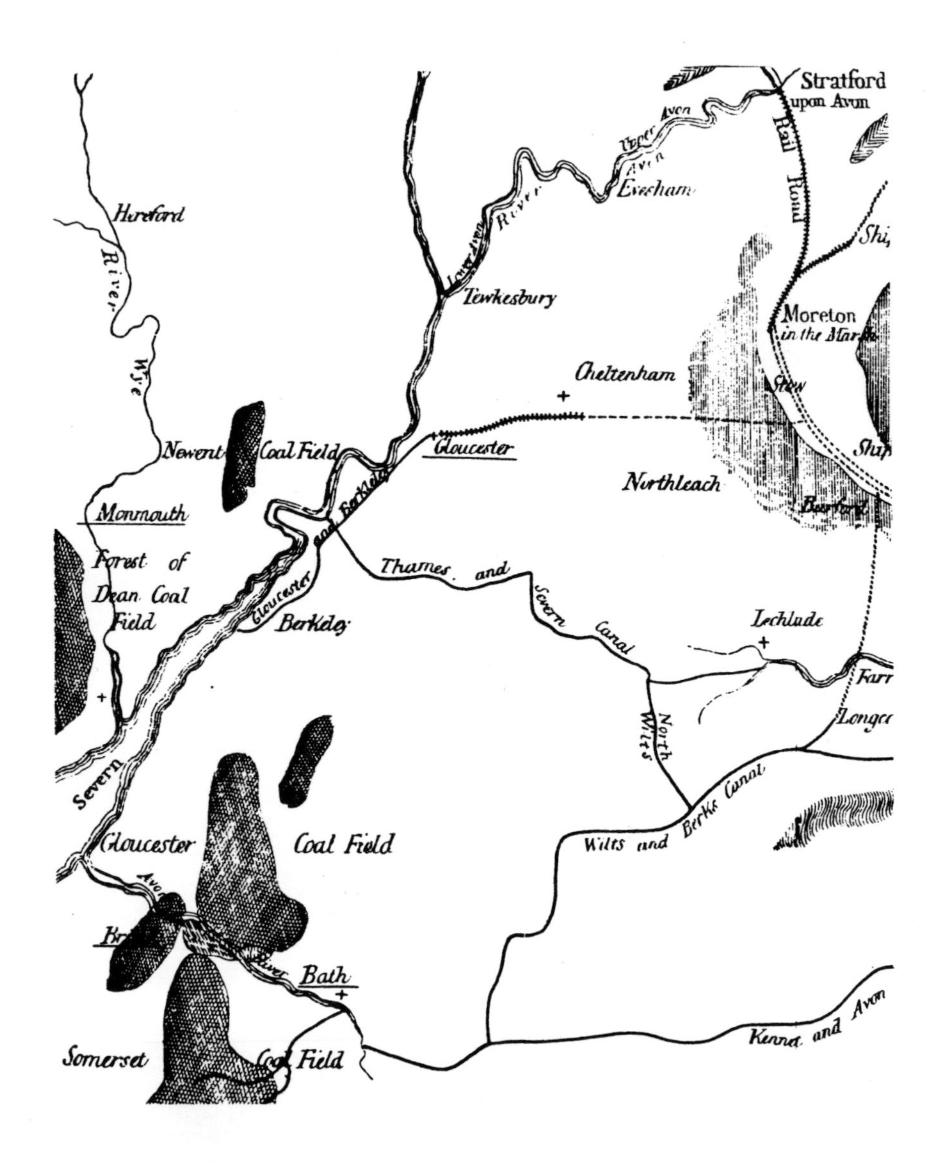

The Newent Coalfield is given prominence in William James' Central Junction Railway prospectus of 1820. Note the proposed extension of the Gloucester & Cheltenham Tramroad to Stow on the Wold.

soon began to speculate in mining on its own account whether or not it had authority to do so. But within a year or two the venture failed and the shareholders also faced the ruinous cost of Oxenhall Tunnel.

For the colliery owners the canal proved a Trojan Horse. It allowed good quality Staffordshire coal to flood the market, henceforth consigning local fuel to the poorer domestic classes, brick-making and lime-burning. However, such is the optimism of miners that efforts continued to bring the works into profit.

As an example of the coalfield's beguiling influence we may cite a proposal put forward in 1820 by the railway visionary, William James. This was to link London with the principal coal-mining regions of England, which on James's plan included Newent. He also proposed extending the Gloucester & Cheltenham Tramroad to a projected main line near Stow on the Wold, but the scheme being years ahead of its time merely labelled James 'stark mad', and came to nothing.[12] James may have had a specific interest in the Newent Coalfield; he was certainly associated with the Gloucester & Cheltenham line,[13] and had earlier opened up the West Bromwich collieries. This connection of Newent with the Black Country is curious, and occurs again and again.

The coalfield rarely employed more than a score of men, the sixty at Newent Colliery, Oxenhall, being the greatest number. Hence there was never a mining community with its own customs and traditions, as often developed elsewhere. Unfortunately, notwithstanding experienced backers, Newent Colliery proved a costly and abject failure and a chapter is devoted to its story.

As to the means of extracting the coal, whether by pillar-and-stall or longwall methods, is not recorded. Probably both were used at different times. Water is the traditional enemy of miners, but except where sandstones overlaid the coal it did not prove a major problem.

Regarding plant and machinery, both for coal and iron ore extraction, with the exception of one colliery the few glimmers of light serve little except to intensify the darkness. We shall see that a steam engine was in course of erection at Hill House Colliery in 1796, but nothing more is known. Its shaft was several hundred feet deep by the 1840s and probably utilised a small high-pressure 'puffer' for winding, thus explaining an engineer who found employment there. At Boulsdon the evidence is merely confusing. One authority implies a steam engine,[14] another affirms a water-engine,[15] whilst a third denies engines altogether.[16] The idea of waterpower is not improbable, a suitable site being in the valley to the east where an ancient leat, perhaps once used for a mill, may still be followed (Grid ref. 712242). A line of flat-rods from a water-wheel could have worked the pumps in the usual way. The ubiquitous horse-whim undoubtedly served, both for raising coal and water. The miners would have gone down in a barrel or else on ladders, using candles for light. Whether the pits were gassy is uncertain, but in any event candles would be used.

Boulsdon and Hill House (Kilcot) Collieries might have justified wagons on cast-iron plateways at surface or underground, which may explain a tantalizing reference in 1912 to 'lengths of small tram rails being found embedded below the surface in different parts of Oxenhall and Gorsley.'[17] Unfortunately no examples have survived. For the industrial archaeologist several sites repay a visit, though

Mining and Geological personalities of the Newent area.
Top left, Sir Roderick Murchison; *top right,* John Phillips; *bottom left,* Thomas Forster Brown; *bottom right,* Frank Evers.

in view of the total extinction of coalfields elsewhere, if nothing of first importance detains him he must be thankful that anything remains at all. The best is lost, fathoms deep, in crushed and flooded galleries that have echoed to no human voice nor ring of steel for over a century. The various surviving features will be detailed later, but interesting and unsuspected legacies are ponds on the outcrop of coal seams. We may suspect in every case they began as small opencasts. Examples may be found west of Knapper's Farm, at Peter's Farm, Hillend on the roadside, near Castle Farm again by the road, and at Castle Tump. Some are now dry or largely filled in, or even, at Boulsdon Croft, turned into a private lawn. At this spot many years ago I was informed of coal dug to feed the boiler of a portable steam engine working nearby. It is also reported that coal occurs in a pond by the roadside between Newent and Huntley.[18]

Finally, for those exploring, legal rights of way including footpaths are marked on the 2 ½ inch Ordnance map SO 62/72 - an invaluable guide. The district has recently been re-surveyed in detail by the British Geological Survey, and by the time these words are in print it is anticipated that both 1:10,000 and 1:50,000 maps will be available. Geological notes to accompany the four 1:10,000 sheets which cover the coalfield are also in course of publication.

Part 1 THE NEWENT COALFIELD

1 Boulsdon

The road from Newent to Cliffords Mesne climbs for much of the way on a ridge between two valleys through which streams run down to the town. At Boulsdon these streams roughly define the limits of extensive old coal-workings, most of which concentrated on high ground on either side of the road.

The first records of mining date from 1608 when the vicar of Newent brought a case of non-payment of tithes against John Beach, bailiff of Boulsdon - coal at that period being treated like a growing crop. Edward Leese of Newent testified to working for a year in a pit in a field called Wheat Croft and 60 loads had been sold in the previous 12 months. Only four to five men were employed. Leese had moved a few years previously from Cannock, a coalmining region in Staffordshire. Two others, Richard Holbrook and William Belloes, came from the Mendips, an ancient area of lead and coal, so it may be that the Boulsdon pits had attracted all three. They left shortly afterwards presumably because the pit closed.[1] The operations were probably sited at the edge of the Coal Measures, close to the road nearly opposite Boulsdon Croft, where in the middle of the last century a field was known as White Croft.[2]

After 1608 mining has left no testimony until the 1760s, by which time Boulsdon for the past few generations had belonged to the Nourses. This was an old and important local family, and Walter Nourse (1654-1742) is still remembered for his manuscript history of Newent. During their reign, pits were operating from 1760-1766 without intermission[3] and no doubt corresponded to the 'Coal Works' marked on Isaac Taylor's map of 1777. Of these ventures it was reported that 'mines were sunk some years since at Boulsdon by the late Mr. Nourse but as the success did not answer the labour and expense they were soon afterwards discontinued.'[4]

For many years the manager of the Nourse estates was John Nourse Morse (1739-1830). He rose to considerable wealth, eventually acquiring the family home, Southerns, in Southend Lane, as well as a great deal of land and property locally. He also ran a general stores in Newent.[5]

In 1789 Morse bought the Boulsdon estate for £3,667,[6] and very soon afterwards sold or leased portions to Edward Hartland and a Mr. Matthews, who I suspect was a son-in-law.[7] Hartland's parents were Thomas and Mary Hartland of Newent and his brother Miles became assistant to the Deputy Surveyor of the Forest of Dean.[8] Miles Hartland was well acquainted with the Forest collieries and we may suspect encouraged Edward Hartland at Boulsdon. In this context Ralph Bigland the historian wrote that 'fresh Coal Pits have been discovered and as a canal from Gloucester to Hereford is now cutting the most Sanguine Hopes are entertained.' The canal was a great incentive and in July 1790 the *Gloucester Journal* reported 'we hear with pleasure that the stratum of coal at Newent exceeds the warmest expectations. The depth from the surface to this stratum is only 41 yards; and though the workmen have sunk six feet eight inches in coal, they have not yet passed through the bed. By means of the intended canal these coals will be constantly and regularly delivered in this city at 8s 6d per ton.'

Two months later it was announced that the proprietors of the Boulsdon coalworks had determined to sink another pit.[9]

In the winter of 1793/4 the canal company decided to open mines on its own account, and applied to landowners for permission to make exploratory borings. Estimates were also sought for erecting 'fire-engines' (steam engines) for pumping. In the course of these endeavours the company approached Edward Hartland of Great Boulsdon 'for leave to raise coal on his land where an 8ft vein had recently been worked'. However the terms were so exhorbitant that attention moved to Kilcot as described later. Having driven too hard a bargain, Hartland then advertised in July 1794 'to be let on lease for 21 years, a very valuable coal work situate at Boulsdon. . . . the state of the coal is between 7 and 8ft deep and at no great distance from the surface.'

Whether the pits were then active may be questioned. The root of the trouble lay in division of ground between Hartland, John Nourse Morse and one or two others, as appears from a pertinent commentary.[10]

> 'The vein of coal lately discovered at Boulsdon was seven feet thick when they left off working. The great obstacle to continuing was the want of an engine to draw off the water. The property in that neighbourhood is divided into small parcels. Coal is probably under the grounds of all the different proprietors thereabouts and should any one person erect a fire-engine he would drain the adjacent grounds as well as his own and would in consequence subject himself to be undersold. To work the pits therefore, either a company should be formed or stipulations entered into by the neighbours to make one common purse for the engine.'

Hartland repeated his advertisement in March 1800, but whether or not as a consequence these sensible proposals were adopted soon afterwards. The Boulsdon Coal Company of which more presently, was behind the revival. It was a Joint Stock company and some of the capital came from Bristol.[11] Hartland and Morse numbered among the investors, Hartland for a long time being the treasurer.

GLOCESTERSHIRE.

TO be LET, for a Term of 21 Years, the extenſive COLLIERY of BOUSDEN, in the Pariſh of Newent, in this County; the Vein of Coal is ſeven Feet in Thickneſs, and lies within 40 Yards of the Surface of the Earth: it is ſituate within a Mile and a Quarter of the Herefordſhire and Glocesterſhire Canal, and about a Mile from Newent.

For further Particulars apply to Mr. Edward Hartland, at Bouſden aforeſaid.

An attempt to let Boulsdon Colliery in March 1800. Glos. Ref. Lib.

Great were the expectations, and William Smith the geologist formed a favourable impression of the prospects. His section of 'Bowsden Coalwork' is reproduced here. According to the Forest of Dean ironmaster David Mushet, a pit was sunk about 1801 to a depth of 80 or 90 yds and a powerful engine erected.[12] A contemporary account referred to 'two coal pits lately sunk at the expense of several subscribers. The depth of the coal is 41 yds; the stratum between 4 and 5 ft, about 7 tons a day are brought up and 12/- per ton is the price at the mouth of the pit.'[13] However, such an output hardly suggests a thriving concern. By 1807 matters had not improved, the collieries being described as still 'in their infant state and not worked deep enough to ascertain either the goodness of the coal or the quantity they are likely to supply.'[14] The same observer continued by admitting disappointment that 'where coal and limestone are in plenty, it does not however appear that any advantage has yet been derived from that source.'[15]

In short the pits had again failed in their promise. Something of the troubles that finally and almost literally, crushed the Boulsdon Coal Company may be glimpsed from a bundle of letters in Gloucester Records Office.[16] They are between Morse, William Deykes and Robert Hughes of Cheltenham, and concern a debt for poles (pit-props). Deykes and Hughes were the Hon. Andrew Foley's agent and solicitor respectively. The letters make pathetic reading with

Great Boulsdon, 1986. This farmhouse was once surrounded by coal-pits, of which little evidence now remains.

Morse, then an old man of 70, pleading for Mr Foley 'to take into consideration our heavy loss and make some abatement,' and excusing delays 'as the concern turned out so extremely bad to the company.'

The sum involved was £144, no small amount in those days, and reflecting the great quantity of timber consumed. Foley eventually reduced the bill to £120 and the debt was settled in May 1811, by which time the pits had almost, if not quite, closed down.

Sir
Being much engaged this week I did not call on Mr Deykes untill this day he says he wrote to you this morning — I have by this Post wrote to Mr Caples recomending to speak to Mr Reuven to agree to accomidate the Business without runing ourselves to further expence — I am in hopes Mr Foley will take into considerat our heavy loss and will make some abatement and I intend to write to him on that head soon as I receive Mr Caples answer I therefore hope you will not proceed — as we are all friends,) untill matters are adjusted I am Sir Your Obedt Servt.

Newent 17 March 1811 JNMorse

I told Mr Deyke I wod write to you

John Nourse Morse's entreaty to Robert Hughes, 17 March 1811. GRO.

The correspondence also sheds light on other partners besides Morse and Hartland. These included John White of Gloucester, Mr Capel and Mr Pruen of Cheltenham, whom Deykes described as 'very obstinate'. He was perhaps Richard Pruen (died 1836) of Pruen and Griffiths, solicitors of that town. It is also tempting to equate Capel with a William Capel who had invested £500 in the newly opened Gloucester and Cheltenham horse-tramroad, and which could have carried Newent coal to the growing spa.[17] White and his brother, who had a grocer's shop, became involved in monies owing to a man named Pitt for hauling props to Boulsdon. Whether because of this or some other cause, White ended up a bankrupt.

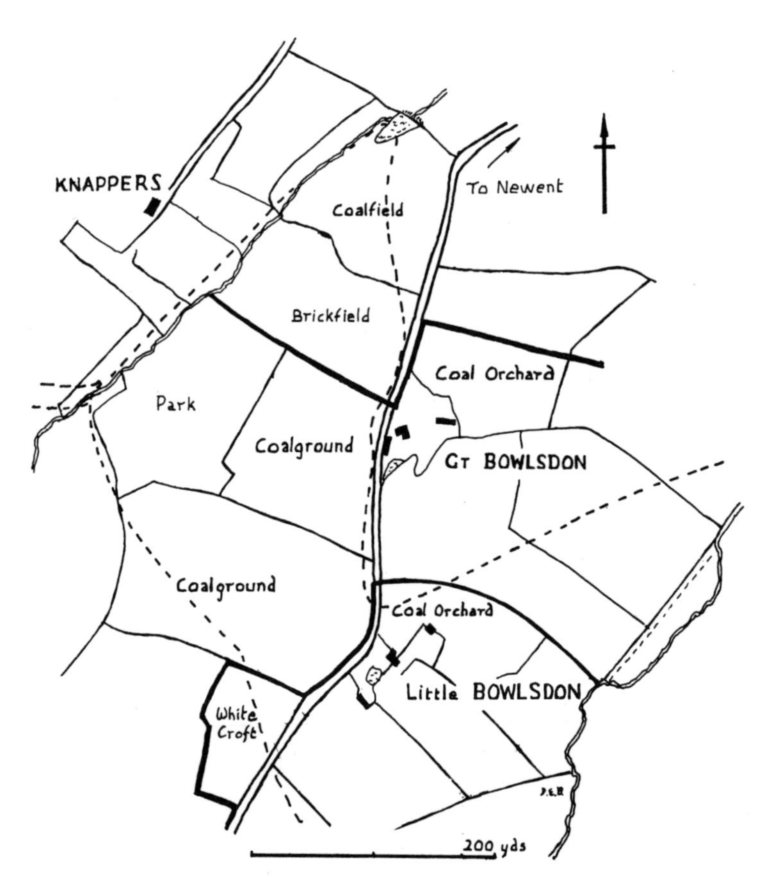

Division of Property GRO.
Based on the 1840 tithe map, the thick lines denote property boundaries.
Top, B. Hook; *centre*, Thomas Morse (probably a grandson of J. N. Morse); *bottom*, John Matthews. The broken line indicates the outcrop of the Coal Measures.

It was a combination of misfortunes that brought the company to ruin. Mushet related that timbers 2ft thick were crushed by the weight of the roof, and there were also claims that bad management, numerous geological faults and flooding had brought about its downfall.[18] Indeed, the potential of the collieries had sunk so low that when the Boulsdon estate came on the market following the death of John Nourse Morse in 1830, no mention of coal was made at all. Furthermore, unlike other parts of the coalfield, there were no landlords here, chained to bankers and desperate for money, continually tempting speculators into a further trial.

As to details of the various workings, the main evidence consists of notes made by the geologist Henry De La Beche after discussions with a miner who had worked in the pits about 1810. From their conversation he made a section of the main pit, as appended here.

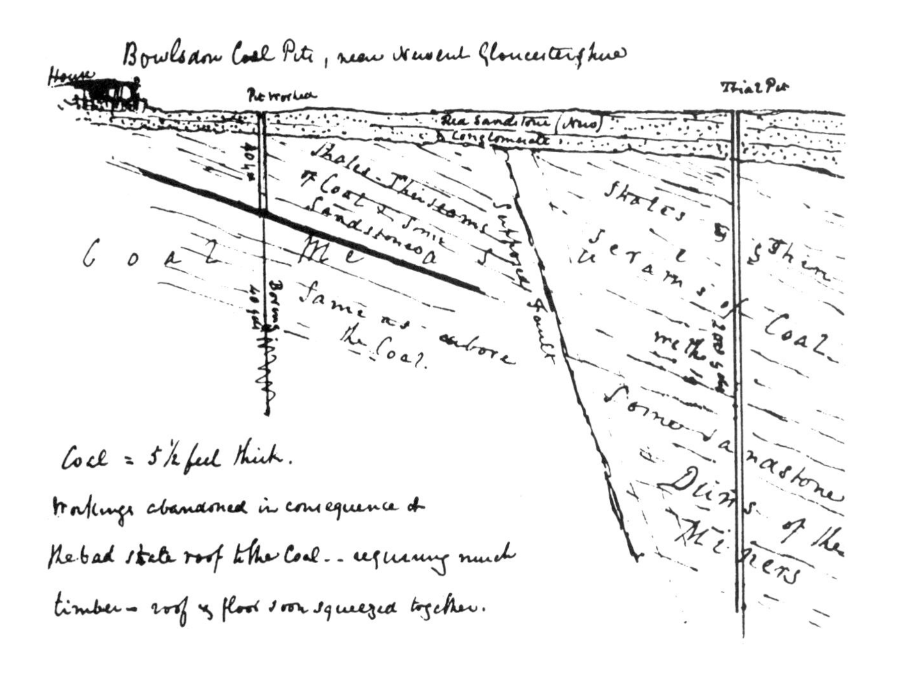

Henry de la Beche's section of Boulsdon Coal Pits. Mem. Geol. Survey.

Bowsden Coalwork 1 M.r S.W. of Newent

	feet & Qrs feet Inc
1 Gravel	2 . 1 . 0 . 0
2 Marl	1 . 0 . 0 . 0
3 Red & White Rock	3 . 1 . 0 . 0
4 Duns &c	8 . 0 . 0 . 0
I Coal	6 feet in to Coals
5 Duns &c	4 . 0 . 1 . 6
II Coal	3/4

Strata ..... 19 . 0 . 1 . 6 Coals 9 . 4

Coals ..... 1 1 . 0 . 4

Total ..... 20 . 1 . 1 . 10

William Smith's section of 'Boulsdon Coalwork', c1802. Bodleian Library.

The shaft (shown as 40 yds to the coal but from his notes only 36 yds) almost certainly equates with *Old Shaft* just behind Great Boulsdon farmhouse as denoted on old 6 inch Ordnance maps. According to the miner the measure dipped at 10 ins/yd. to the east. A boring put down 40 yds below the coal revealed no significant seams, and to the east a trial shaft sunk 200 yds[19] found nothing workable.[20] An intervening fault of large throw was deduced. A further piece of evidence is from Mushet who recorded that before the pits closed a level was driven across the rising plane of the measures (i.e. westwards) and proved the existence of lower coals. Perhaps these latter conformed to the 3ft seam of Smith, which surprisingly was not noted by the other authorities. In 1879 it was recorded that Boulsdon could claim no less than three separate seams - 4ft, 3ft and 6ft in thickness.[21] Unfortunately no further details were given.

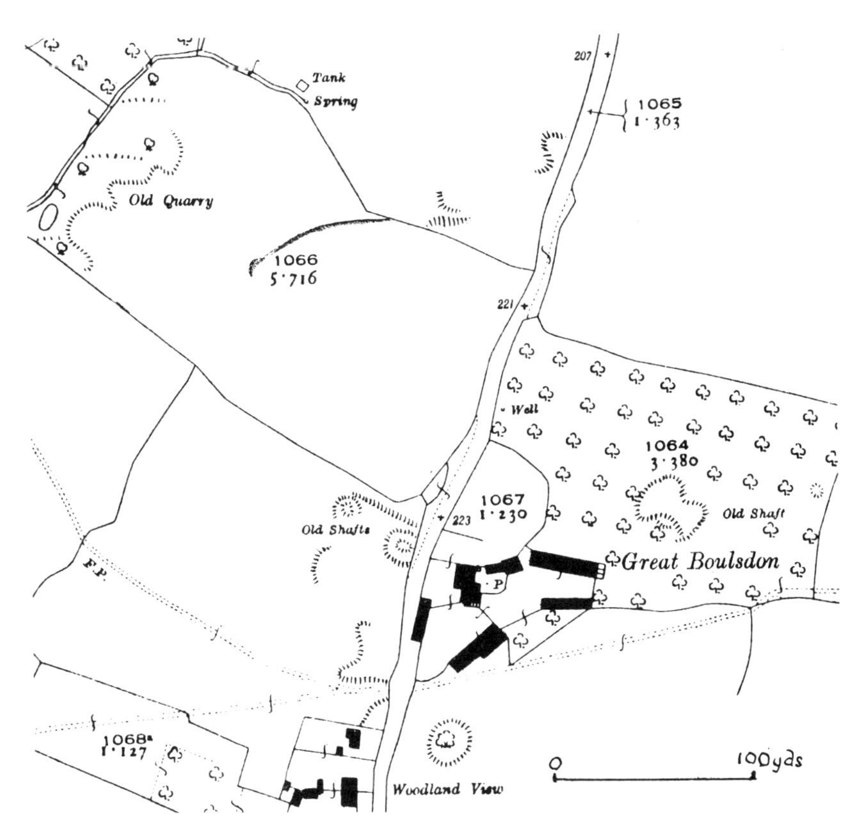

Great Boulsdon in 1923 25 inch Ordnance Survey
The main workings were behind the house. The Spring (top) probably denotes an adit or water-level for drainage.

Coal recently revealed in the bed of a pond east of Great Boulsdon.

It is a great misfortune that so much information once available on the extensive working around Boulsdon has been lost forever. It is now impossible to identify the sites of the various shafts which must have numbered well into double figures,[22] or, with the exceptions noted above, to tie them in with the various sources as summarised below:

Source	*Thickness of Seam*	*Depth to Coal*
Gloucester Journal	6ft 8ins	41yds
Mushet	6ft	80-90yds
Smitn[23]	6ft in four beds	30yds
Rudge	4-5ft	41yds
Murchison	5ft 8ins, in four beds (1ft 6ins, 10ins, 10ins, 2ft 6ins)	36-80yds
De La Beche	5ft 6ins	36yds

The Old Series One-inch Geological map shows the main seam outcropping in a line running NNW from Little Boulsdon (Boulsdon Croft) to Knapper's Farm.

The pre-19th Century workings were no doubt west of the road, except for those near Boulsdon Croft where the coal is shallower. The later pits were sunk deeper and to the east, mainly in the vicinity of Great Boulsdon under a mantle of Triassic sandstone.

In recent years coal was found in excavating a pond in the valley to the east. Its relationship to the main workings is uncertain, and may represent a seam hitherto unrecorded.

After an interlude of nearly two centuries little remains of coalmining at Boulsdon. Almost opposite Hope Cottage a grassy mound surmounted by a small shed denotes a shaft, and in ploughed fields below the road extensive patches of blue and yellow clay with dark areas of coal-smut and pieces of coal testify to bell-pits or open-workings. Similar evidence just west of the road 200 yds north of Great Boulsdon may indicate the abortive shaft mentioned by De La Beche.

Finally, low down in the same field a spring emerges from what was probably the adit or water-level draining the pits behind the house. Towards Common Fields is an old earth dam, long since breached, but whether having associations with mining is unknown.

Between Boulsdon and Kilcot, Coal Measures outcrop in a narrow band, but the witness of a trial in the ploughing about 300yds SSE of Wyatt's Farm is the only evidence of activity. Probably seams lie concealed under sandstones to the east.

2 Kilcot

Hidden between the Kilcot Inn and the lane from Oxenhall to Gorsley, Coal Measures have been extensively worked and not without a degree of success. The collieries were important enough to merit inclusion on two maps published in 1817 and 1831 respectively - Henry Price's *Herefordshire* and the Old Series Ordnance.[1]

This small area lies on the borders of Newent and Oxenhall parishes, but in the times we are considering much of it formed a detached part of Pauntley. Mining may be inferred quite as early as at Boulsdon, for in 1608 John Hodges of Kilcot was described as 'a coleminer' about 40 years old and of a middle stature 'fit to make a musketeer'.[2]

Kilcot Estate was acquired by Walter Nourse on wedding the daughter of John Bourne of the Conigree. It eventually passed by marriage to John Lewis of Llantilio Crossenny near Abergavenny, who sold it to John Nourse Morse about 1783.[3] A few years later Morse disposed of Hill House Farm to a Mr Wood of Newent,[4] Hill House Colliery being afterwards sunk in one of its fields.

The ground between White House, Hill House and Lower House was intensively prospected, and in the 1870s men were alive who could recall four or five shafts in one meadow. In 1792 Bigland recorded that 'on an Estate called the Lower House Farm belonging to Messrs. Phillips and Wood, coal is working with a good Prospect.'

It was to this spot that the Hereford & Gloucester Canal Company turned in April 1794 and their records have yielded much information.[6] After negotiating with Phillips a shaft was sunk, but being very near previous workings and on the crop of the seam or vein, the coal proved fit only for lime-burning. The committee in its annual report made the best of a bad job by adding 'for which purpose it was found superior to any which had before been used. Another pit is now sinking further from the old works and more on the dip of the vein.' To reduce costs the shaft lining was to be secured 'in timber and not bricks, and a very particular account taken of the expense'. Terms of $^{1}/_{7}$th were agreed as from 24th October 1794 on the price at the pit-head - a high royalty bearing in mind that the venture was not a proven proposition.

Early in 1795 the Rev. Foley of Newent purchased ten waggon loads for the poor, and 'at the colliery they are got down to a vein of coal somewhat more than 6ft in thickness and this only 31 yards from the surface'. A 'railroad or small collateral cut' was then considered to the canal a mile away, with a possible extension in the opposite direction to limestone quarries on Gorsley Common

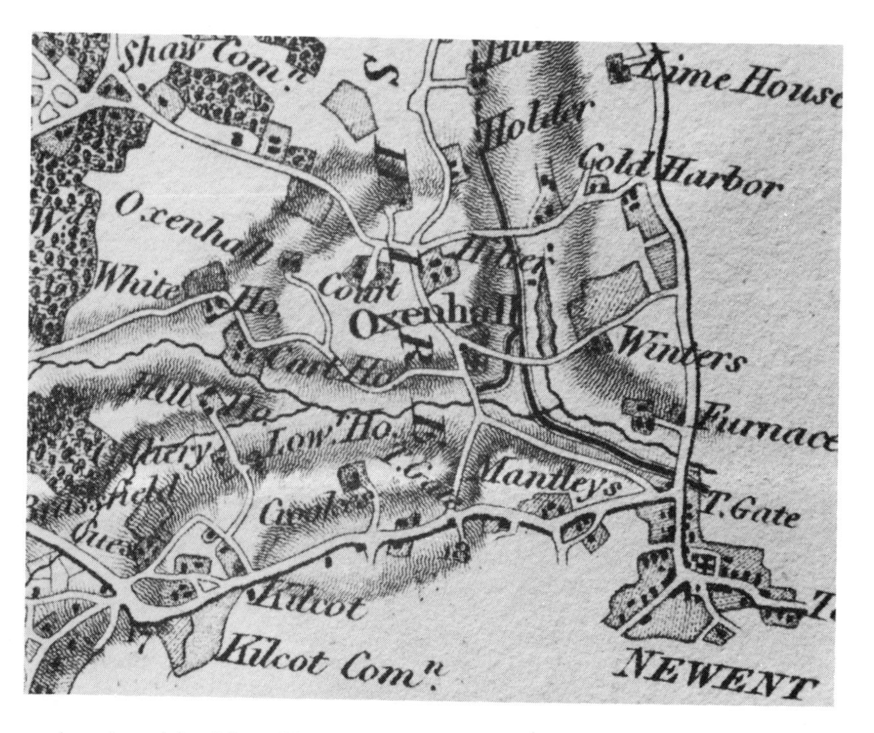

An enlarged detail from Henry Price's map of Herefordshire, showing Lower House Colliery about 1797.

-then a wild open tract, uninhabited except for squatters and vagrants. Lime from this source was highly regarded as a manure,[7] and five kilns which still survive in a precarious state in Green's Quarry are a forgotten legacy of this once important local industry.[8]

However, the canal from Gloucester had yet to open, and coal went to the city by road. The pit-head price of only 9/- per waggon load of 2½ tons probably reflected an indifferent quality, though the rate soon increased to 14/- for two tons 'to whoever will fetch it'.

The venture still being far from profit, in a further attempt to find better coal the company made an agreement in May 1795 with William Wood of Newent to sink a pit on land at Hill House, between the present pit and the Ell Brook. Meanwhile one Robert Miles contracted to haul 200 tons from the colliery to the canal basin, now open at Newent, for 2/6d per ton. He received £100 on account and a further £80 afterwards. In the summer of 1795 it was ordered that Joseph Phelps a tenant of Samuel Beale of Upton on Severn, should be compensated for losses to a crop of barley in ground belonging to the colliery. To prevent further trespass the gate was locked and a notice erected which read *The Key to be had of Mr Webster at the Colliery.* Another key was given to Miles.

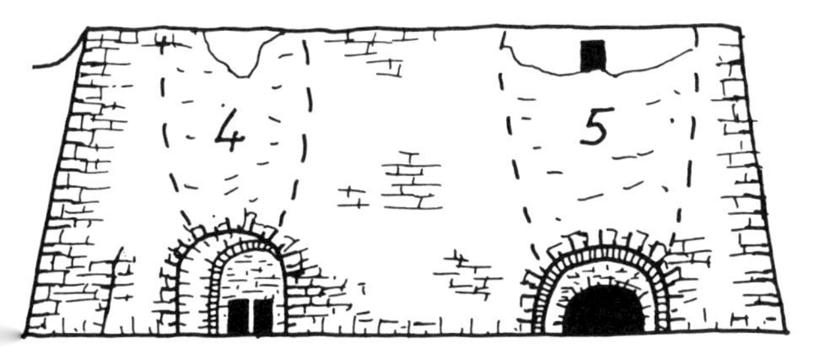

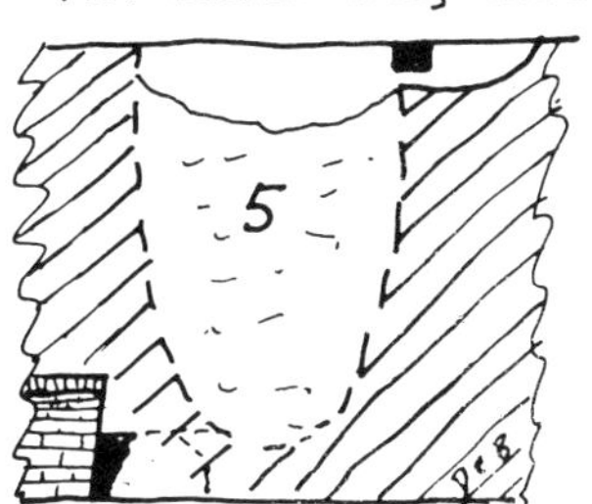

This battery of limekilns in Green's Quarry, Gorsley, is now in a ruinous state. Nos 1, 2 and 3 are 18th century. 25 yards away, Nos 4 and 5 are much larger and perhaps 19th century. No two kilns are the same, and the tunnels **a** and **b** (now largely filled with rubble) are difficult to explain.

Note: kilns 3 and 4 were summarily demolished as these pages are passing through the press - a sad loss to the local heritage.

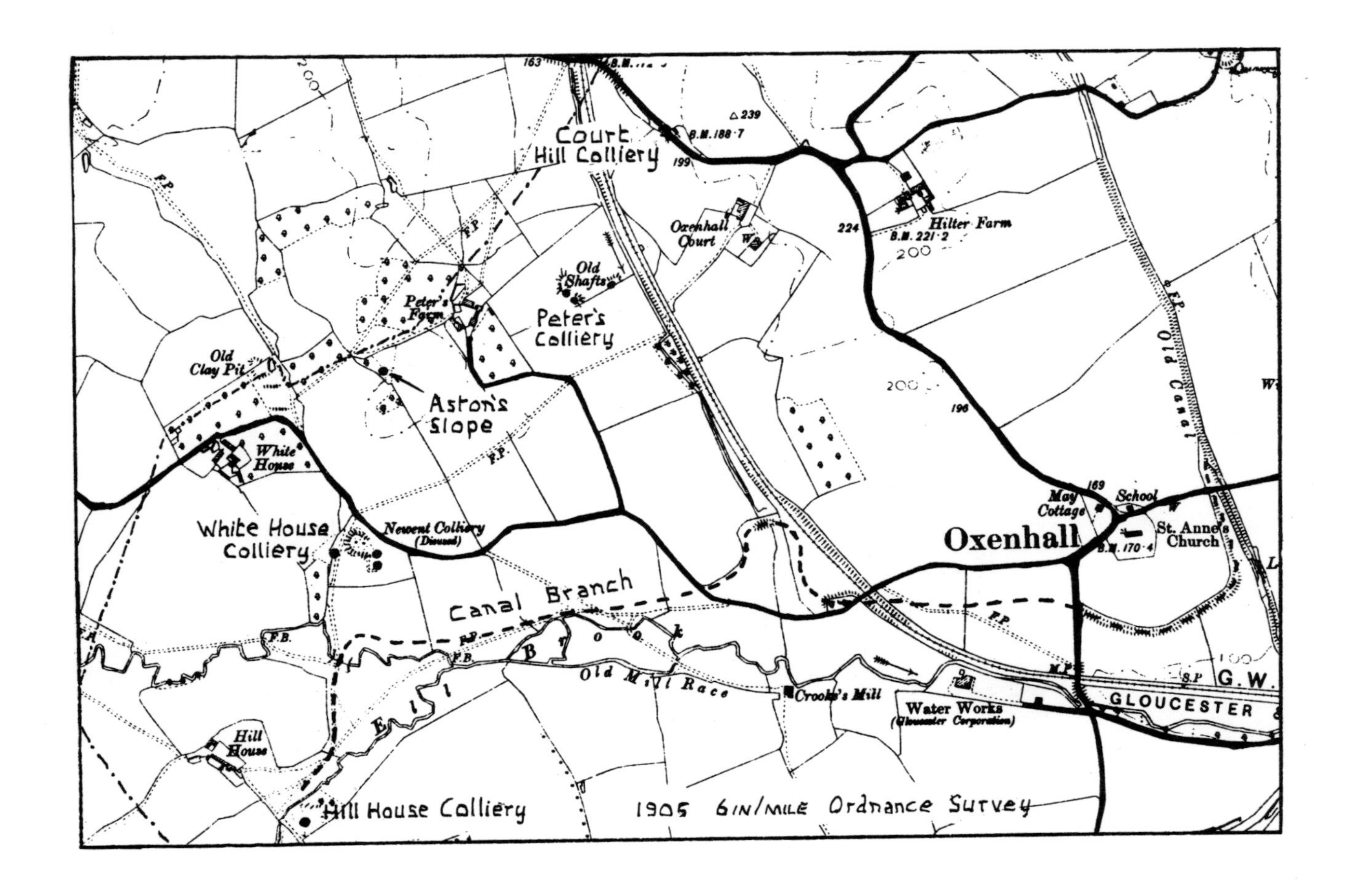
Court Hill Colliery
△239
B.M. 188·7
199
163
Oxenhall Court
224
Hilter Farm
B.M. 221·2
200
Old Shafts
Peter's Farm
Peter's Colliery
Old Clay Pit
Aston's Slope
White House
F.P.
198
Old Canal
May Cottage
169
School
St. Anne's Church
B.M. 170·4
Oxenhall
White House Colliery
Newent Colliery (Disused)
Canal Branch
F.B.
Brook
Old Mill Race
Crooke's Mill
Water Works
(Gloucester Corporation)
M.P.
S.P.
G.W.
GLOUCESTER
Hill House
Hill House Colliery
1905 6IN/MILE Ordnance Survey

But matters such as this were trivial compared to imminent changes. One of the men drawn to the coalfield was Richard Perkins (1753-1821) a surgeon from Oakhill on the edge of the Somerset Coalfield. In the company of William Smith and Samborne Palmer a colliery owner, he had lately returned from a tour of coal-mining regions on behalf of the Somerset Coal Company. Whether Perkins' interest in these embryo Gloucestershire pits arose from any persuasion by his companions we do not know. On 1st July 1796 he took over the Lower House and Hill House Colliery leases as a going concern.

It was agreed to build a branch one mile long from the main canal near Oxenhall Church free of charge provided he pay 2d per ton for coal passing over it. Perkins was also required to supply 4,000 tons per annum for burning bricks and lime at 4/- per ton for rubble coal and 3/6d for slack, until the main canal was completed. In addition he agreed 'to deliver on the banks of the branch and pass thereon 35 tons per day[9] for the first three months it shall be open and 70 tons per day afterwards', provided such a quantity could be raised and a market found. The prices exceeded those at Forest pits which in 1788 averaged as follows:[10] House coal 4/3d per ton, Smith's coal 2/6d, Lime coal 1/6d. Cartage to Newent would have added several shillings, so it appears the Hill House figures simply reflected what the market would bear. At this period the company faced desperate trouble with Oxenhall Tunnel and these terms, so favourable to the purchaser, revealed its anxiety to shed an ailing speculation at almost any cost.

On gaining control Perkins introduced changes in methods of working, and began by dismissing the miners. John Webster, the foreman, and his family were despatched to their home in Nailsea, a small coalfield beyond Bristol, and allowed four guineas for the journey to be paid on arrival. But as regards coal traffic, prospects began to look increasingly gloomy with the canal committee darkly reminding the owner of his obligations. 'Get on with sinking your new shafts,' it wrote, 'as a disappointment in this instance will be big with consequences the most disagreeable to yourself and the company.'

In the November of 1796 shareholders were told of Perkins' experience in mining matters, and that he was erecting a steam engine. They were informed that this, together with his substantial capital investment 'justifies the sanguine hopes long entertained by your committee, that it will be ultimately productive of great pecuniary advantages to your undertaking.' It was a forlorn hope. Nonetheless the company clung to the branch for decades in hopes of better days. Reading between the lines, most of Perkins' output went elsewhere with substantial quantities to the limekilns at Gorsley. At the end of 1799 he was raising considerable tonnages, some of which got as far as the Stroud area for the use of clothiers. After 1800 the bankrupt and dispirited canal company descended into a state of suspended animation from which it did not recover for many years, and further gleanings must be sought elsewhere.

Collieries in the Kilcot-Oxenhall area, and the abandoned 1790s canal branch to Hill House Colliery. The chain-dotted line marks the western boundary of the Coal Measures.

A public footpath crosses this 18th century aqueduct carrying the canal branch over the Brockmorehead Brook near Hill House Colliery. It is endangered by proposals for a new route for the A40 trunk road.

It seems likely that Kilcot pits closed early in the new century, for Rudge, writing in 1803, refers only to Boulsdon. And when Richard Perkins advertised Hill House Farm for sale at the Bull Inn, Newent, in 1808, there was no mention of coal.[11] Nevertheless, though it was later claimed that colliery speculations had ruined him,[12] this does not seem to be so, for he later emerged deeply involved in coal-mining in South Wales.

As for Hill House Colliery it had revived by August 1839, an event which the canal company hoped might justify re-opening the branch. Details of those employed and residing locally in 1841 are revealed in the Census Returns as below.

Address	*Name*	*Occupation*	*Age*
Kilcot	William Perkes	Engineer	20
	Robert Rudge	Coal Miner	20
	William Predett	Coal Miner	40
Near Lower House	John Weale	Coal Miner	35
	Elizabeth Weale	-	40
Colliery Cottage	Joseph George	Coal Merchant	40
Collier's Cabin	John Day	Coal Miner	50
	Henry Portlock	Coal Miner	15

The Weales had seven children between one month and 15 years, several apparently also being miners.

Hill House Colliery in 1970. *Top*, looking west; *bottom*, looking east, showing the grassgrown humps and mounds of the old tips. The site has since been completely levelled and converted to arable purposes.

John Phillips implies that the pits had closed by 1842,[13] but it may not be so, for in 1846 slight workings were still carried on near Hill House.[14] In 1857 Hill House Farm[15] together with Oxenhall Court was sold to R. F. Onslow a prominent local landowner, by order of the Court of Chancery, but not for a number of years did a resurgence of interest arise in the Onslow Estate minerals. It came about in the early 1870s as the result of a boom in the coal trade. One of the consequences was the following promotional report in the *Mining Journal*,[16] which throws considerable light on the previous operations.

> 'Mr William Perks who was engineer at the Hill House Colliery 30 years ago (1843) writes that they had a pit there about 100 yards deep, and three veins of coal - the top vein about 1ft thick; another 1ft 8in, and in the bottom vein (wrought by them) 6ft, with a good rock top requiring but little timbering . . . they were not permitted to go where the best coal was, as the property belonged to different parties . . . He recommends sinking a pit near the mill[17] where they will have an area of some hundreds of acres. The party he was connected with had only a few acres, yet their take lasted for 12 years, during which time they were so little troubled with water that they drew it all with a barrel. The Hill House pits were worked by Mr Perkins in 1796 and he realized a fortune out of them. . . .'

Perkes' inference that the pits operated for 12 years at a stretch may be questioned, but it does seem that they achieved a measure of success apparently by sinking deeper to a seam beneath the old workings.

The promotional attempt had a sequel, for according to the agent of the Onslow Estate a shaft was sunk at Lower House.[18] When W. S. Symonds led the Cotteswold Naturalists Field Club there in May 1874 he recorded 'visited Hill House Colliery. The workings appeared to be on a very small scale but were probably quite commensurate with the importance of the venture.'[19] Tea was afterwards taken at Stardens by courtesy of R. F. Onslow.

This was pretty certainly the last of mining at Kilcot and may correspond with a windlass over an open shaft, still evident about 1900.[20] As to the final cause of closure, in 1885 a mining engineer engaged to consider the desirability of continuing a borehole on the Onslow Estate, reported that 'a large fault stopped the workings at Hill House Colliery and brought in the New Red Sandstone . . . the seam of coal was 8ft thick and I think the strata was dipping rapidly.'[21]

As to details and extent of the underground workings we are hardly better informed than at Boulsdon, though boreholes drilled by the British Geological Survey within a few yards of Hill House Colliery shaft have given useful evidence.[22]

In the 1880s a South Wales mining engineer, Thomas Forster Brown, had access to the plans of Hill House, but their subsequent whereabouts as well as those of a copy said to exist locally, have never been traced. For the early years certain facts have survived from the accounts of visitors, the first being an employee of the Ordnance Survey, Henry Maclauchlan; 'the seven ft coal was the one principally worked; but it was much disturbed and dipped rapidly at two ft in a yard to the N.E. or E. In 300 yards four faults of considerable magnitude were encountered and the coal contained a large quantity of sulphur.'[23] If the coal-face extended for 300yds we may presume the workings ranged underground roughly from Lower House to Hill House.

Soon after Maclauchlan followed Roderick Murchison, whose field notebooks make an illuminating commentary:

> *Lower House* Pits were sunk between 40 & 45 yards through the new Red Sandstone to the coal and one was 50 yards depth. In this spot and at Hill House Colliery it is evident that the coal must soon have been cut out by the old red sandstone. . . . At Lower House the coal in one sinking dipped North East, in another contiguous, South East about 25°. These opposite dips were produced by a fault of zig-zag direction and which occasioned an upcast of the coal of 25 yards. The fault was filled with white sandstone.
>
> In sinking these shafts new red sandstone 6-7 yards was first sunk through, then
>
> | | |
> |---|---|
> | Whitish clay | 21ft |
> | Hard whitish sandstone with plants | } 75ft |
> | Brown reddish marl | |
> | Clunch | |
> | Coal | 7ft |
>
> From the top of this coal to the base of the coal-bearing strata, about 23ft, including 4 coals viz.
>
> 7ft coal
> 2ft coal
> 1ft 6in. coal
> 1ft 2in. coal
>
> The bottom a sulphurous coal.

In another account Murchison remarks: 'At the Hill House works undertaken by Squire Perkins the coal pitched very sharp Eastwards, viz. 14 inches per yard and could not therefore be followed in the dip - also attended by innumerable faults.'

Murchison, who stayed at the George Hotel, also found time to scribble down personal experiences; his encounter with Thomas Phillips in 1833 adds a tantalizing insight into the human side of the Kilcot pits.

> 'This miner had expended £500 . . . in these collieries from which he and his [word illegible] were expelled by an outburst of water, which not only filled the works but actually rushed to the top of the 50 yard shaft and overflowed the adjoining meadow.
>
> Mr Phillips who has been printer, auctioneer, land surveyor and coal speculator . . . was an admirable specimen of an ingenious obliging restless spirit, a true speculative Englishman. His father, who had possessed some little land around this spot was now reduced to a cottager in his 80th year . . . he is bent upon America next year and says he can get £500 together for the start. He offered me a good draught of cyder . . .' [the rest is illegible.]

Details of the colliery in Perkes' time are equally obscure, though something may be deduced from the BGS boreholes. The appended section shows a tentative and very basic interpretation which assumes one major fault only. By sinking deeper, coal under the old workings could have been exploited, as well as the old seams beyond the fault, by means of a cross-measures drift to the north-east. The Pauntley tithe map throws useful light on the period. The boundary is shown between Hill House and Lower House properties, the latter being owned by Archdeacon Onslow (father of R. F. Onslow) and leased by George Hook. Hill House then belonged to Samuel Beale of Upton-on-Severn, and that part north of the Ell Brook was leased to Edward Dowell. The tenant between the stream and Lower House estate was Joseph George, coal merchant, who lived at Colliery Cottage, no 315 on the map.[24] Hill House Colliery occupied the ground 314, described as 'coal pits and piece', and we cannot help wondering whether

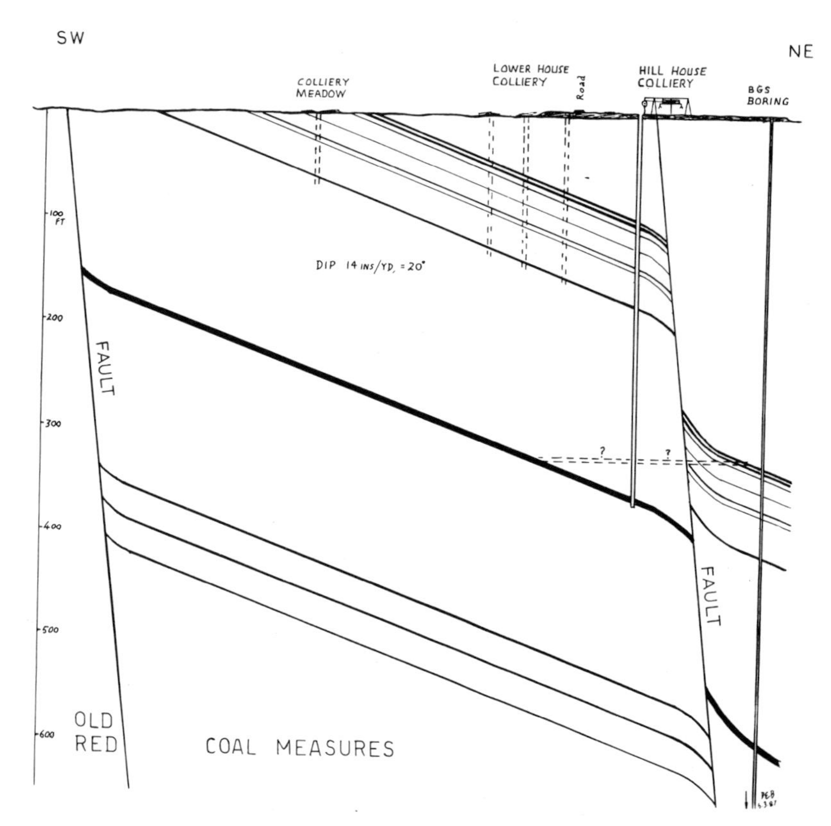

A tentative section through the Kilcot pits, as far as can be deduced from records and borings. Faulting is much more complex than indicated here.

George's influence had re-opened it not long before. Whether any pits were sunk on the opposite bank of the stream is not known, neither is it certain whether Hill House Colliery employed at all times only a single shaft.

The Lower House pits, marked as 'Colliery' on Price's map of 1817, occupied the ground 316 'part of coalfield'. They probably also encroached on fields 313, 317 and 881 which either reveal Coal Measures at surface or else an overlying mantle of New Red Sandstone.

Colliery Cottage was demolished before 1884, and the extensive hummocky tips of Hill House Colliery were bull-dozed a few years ago. A grassy tump survives in 336, possibly from a shaft or dipple to the south. Close by is a long shallow depression striking towards Kilcot Hill, perhaps the vestiges of an opencast. This field, 336, is still known as Colliery Meadow,[25] probably in deference to workings of which all records are lost.

These scarcely discernible features in quiet fields bordering the Ell Brook are the sole memorial to the Kilcot pits that persevered on and off, for generations.

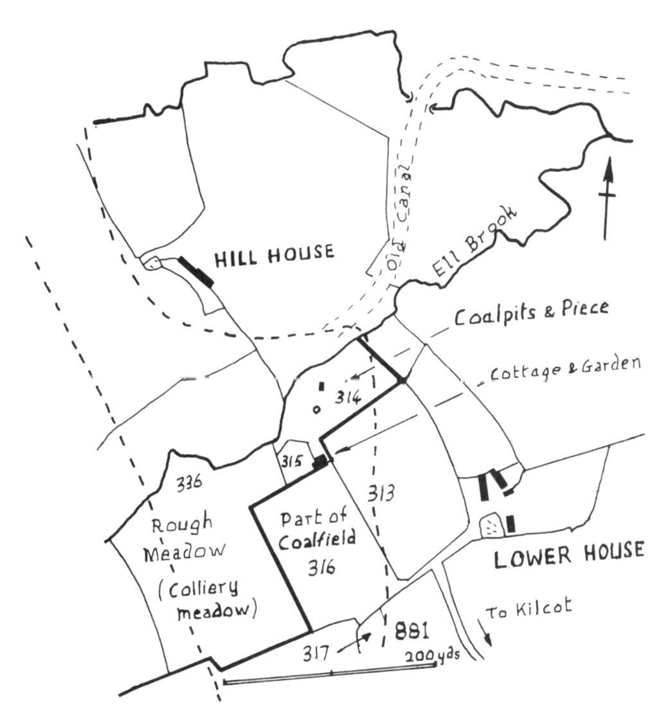

Pauntley Tithe Map, 1840 GRO.
Field names are shown, and the thick line indicates the boundary between Hill House and Lower House properties.

3 Oxenhall

Northwards from Hill House, borings levels and shafts have explored the ground the whole distance to the parish boundary at Castle Tump, and coal and iron ore raised at various places. According to a report in 1879,[1] some spots had been successfully worked but abandoned after only a short trial:

> 'Want of capital had been generally assigned by the workers as their reason for leaving the experiment . . . but well-informed folk declared that the Forest of Dean Coal-owners, dreading the effects of a near rival district, resorted to bribery, and to this was to be attributed the sudden and repeated abandonment of an enterprise which in its beginnings promised every success. So largely was bribery alleged that an old man who had helped to work one of the collieries had been heard to say that when the shafts were covered in, 300 tons of coal were left quarried in the workings.'

The history of mining in Oxenhall was largely influenced by the Foley and Onslow families about whom a few words must be said. The Foleys dominated the West Midlands iron industry for well over a century, the dynasty being founded by Richard Foley, a nail-maker of Stourbridge. His son Thomas (1617-1679) of Great Witley in Worcestershire, eventually acquired much of Newent and Oxenhall parishes, also a number of charcoal blast-furnaces in West Gloucestershire of which one, generally known as Elmbridge Furnace, stood just north of the town. The works comprised various buildings including a steelworks and a fine charcoal store which probably had much earlier origins.[2] The latter and the furnace blowing-house are now Listed Monuments, and the road to the site is still called Furnace Lane. Pig-iron was the main product, but homely items such as firebacks were cast,[3] and a number remain in the locality. More on Newent Ironworks is given in a later chapter.

In the 18th century Thomas Foley the third owned practically the whole of Oxenhall and was created a Baron in 1776. His youngest son Andrew inherited the estate. Subsequently it passed to Richard Francis Onslow by marriage and thence to his son Richard Foley Onslow, born in 1803. In the same year the father was appointed vicar of Newent, a position he retained for over fifty years.

Whereas the Foleys had been absentee landlords, Richard Foley Onslow lived locally. He continued the traditions of a country gentleman, conscious of his duties but at the same time eager to preserve the image of a fox-hunting squire. Unfortunately the cost of enlarging and maintaining his home at Stardens together with various other outgoings landed the estate in mortgage to the tune of £46,000 at the time of his death in 1879. Aided by a complicated will and two codicils the property proved a rich harvest for bankers and lawyers alike, and we may wonder how much remained for beneficiaries.[4]

Throughout most of the Foley and Onslow reigns determined efforts were made to exploit the mineral potential of Oxenhall, the earliest documentary reference so far discovered being in 1752 when the Foley archives[5] record that William Gatfield received £1.10.0d for '31 days Boreing for Cole'. 'Several workmen boreing for Cole' were paid £11.6.4d.

This fireback measures 26ins × 51ins and was found in the rubble of a demolished cottage in Bradford's Lane, Newent. It can now be seen at the Glasshouse Inn and almost certainly came from Newent Ironworks.

Nearly 40 years later the proposed canal gave new impetus, it being reported that 'great exertions are now making to establish collieries on the estates of the Hon. Andrew Foley at Oxenhall.'[6] These were probably associated with a borehole already begun under the direction of William Deykes, Foley's agent. By 23rd January 1791 it had reached a depth of 71ft, passing through five seams totalling nearly 8ft in thickness.[7] A section of the borehole is given in an appendix. Work on Oxenhall Tunnel began a year or two later, and a local observer recorded that within a few yards of its projected course different thin strata of coal had been pierced through, 'in the whole area about 3 feet thick within the depth of 20 or 25 yards. The stone of the hill seems also strongly impregnated with iron-ore.'[8] This description suggests a further boring, between Hillend and Castle Farm.

In 1794 a man who with Richard Perkins was to make a considerable impact on development of the coalfield, became Lord of the Manor of Dymock. He was John Moggridge (1731-1803) a clothier from Bradford-on-Avon, prior to moving to Boyce Court, near Dymock.[9] In the same year, 1794, his sister Elizabeth married Perkins' son, also named Richard. Moggridge had an only son John Hodder Moggridge and the four men became involved in a commercial venture known as Perkins, Moggridge & Co. In allusion to their enterprise, on completion of the canal to Ledbury in 1798 the *Gloucester Journal* was able to report 'it has given occasion to the opening of valuable Coal Mines long neglected, for want of water carriage, which are now working on an extensive scale and yielding abundance of excellent coal.'

Although we may take the announcement with a pinch of salt there is no doubt that whilst construction of the canal sustained a demand for fuel and materials, Perkins, Moggridge & Co., carried on a substantial trade. J. H. Moggridge found time for involvement in public life and in 1809 was High Sheriff of Gloucestershire. But by now the limitations of the Newent Coalfield were clear. In 1809/10 Moggridge and Perkins sold much of their estates in the area, Moggridge and

Perkins junior (1772-1850) transferring interests to Penmain Colliery[10] in the heart of the South Wales Coalfield. Near Llanrumney Moggridge built a mansion called Woodfield and founded the town of Blackwood as a means of inproving the housing and well-being of his miners.[11] The records of these adventurers during their Gloucestershire interlude are a key to the history of the Newent Coalfield, and perhaps they may yet come to light.[12]

During the next century we may suspect that few decades passed without attempts on the Foley or Onslow estates. In 1858 the iron deposits above the Coal Measures were described in a book on mineralogy as occurring 'in regular beds of considerable extent . . . and usually at the base of the New Red'. For the years 1855-65 the *Mineral Statistics* list a mine under the name Oxenhall Colliery, the site of which is uncertain, and in 1864 Onslow applied to the canal company for reduction in rates for iron ore shipped from Newent.[13] This interest in iron continued, and aided by a boom in trade, in 1873 promoters named Aston and Brown projected a company for working the Onslow estate of over 3,000 acres at a dead rent of £100 per annum. The royalties were 1/6d per ton on iron ore and 4½d on coal. After spending £600 on trials, coal and iron were traced for over 1½ miles and pits sunk 6 or 7 yards deep.[14] This led to considerable developments which will be described presently.

Mr MOGGRIDGE'S of the BOYCE near DYMOCK

An illustration of Boyce Court in 1807, the home of John Moggridge.

Even a decade later the coal and iron deposits were not quite dismissed. When the G.W.R. built the Gloucester-Ledbury line it levelled a large triangular area near the Oxenhall-Gorsley road, so it is said, for sidings should the need arise. Ironically the canal branch built some 80 years earlier bordered this very piece of ground, and from the old railway bridge in the lengthening shadows of an evening sun may be discerned the remnants of these vain endeavours.

Richard Foley Onslow's death in 1879 and the disastrous failure of Newent Colliery, (see next chapter) did nothing to quell a determination to make something of the minerals. The trustees engaged a consultant whose name was Thomas Forster Brown, one of a family of distinguished mining engineers and Deputy Gaveller of the Forest of Dean. Early in 1882 Brown went over the ground with Capt. Onslow and his agent, a Mr Greenwell, Charles Jones the blacksmith of Hillend, and Maule, the Onslow's solicitor from Newham. A report of the excursion is preserved in the Gaveller's Office, Coleford.[15] It will be much quoted, though Brown's reluctance to observe rules of punctuation render it in places a rather ambiguous source.

Turning now to a more particular account, we will begin at White House.

White House

Coal Measures outcrop in a band sweeping round behind the buildings which stand on a knoll of sandstone. In the courtyard a well is sunk 87ft and no doubt ends in Coal Measures. Some years ago Mr Farnham the owner showed me a curious brick-lined chamber or shaft which enlarged as it went down; somewhat bell-shaped as I remember. It was covered with a slab near the hedge 100-150yds SE of the house; the site is now lost. The area has witnessed many trials. About 1800 coal was found in a shaft about 60ft deep, 150yds SW of the house, and in the 1870s Aston proved a 2ft seam beneath shale at the roadside.

To the SE is Newent Colliery, marked by a substantial grass-grown tip, and in the valley between was the enigmatic White House Colliery, about which little more is known than its name. We have it on the authority of the Cheltenham geologist Linsdall Richardson that a boring was made and a shaft sunk by the Newent Colliery Co. on this site, which no doubt corresponds to *Trial Shaft* marked on the 25 inch map of 1884.

These events were recorded in the *Gloucester Journal* during the summer of 1877.[17] 'Some time ago the work of sinking a shaft was begun and early this week coal was struck at a depth of more than 100 yards. The seams passed through have a thickness of about 6 feet and the boring is being continued . . .' A fortnight later further particulars were furnished. 'After passing through 8 feet of iron ore and 16 feet of clay, a seam of coal 6 to 8 feet thick was struck beneath which a second seam was found from 2 to 3 feet thick. Boring is still being continued in different parts.' A remarkable survival from the period takes the form of a wooden plank inscribed with a section of strata and accompanied by coloured panels alongside. We may suspect its original purpose was to grace a shareholders' meeting, though the reason behind so novel a presentation is hard to conceive. It finally came to rest in a solicitor's office in Newent where it has resided ever since, and I am indebted to Mr Paul Eward for permission to reproduce the accompanying photograph. According to Richardson the plank records the actual log of the White House Colliery boring, and there is no reason to doubt it. The inscription is given below:

SECTION OF BOREING

	yds	*ft*	*ins*
Red Sandstone Rock			6
Red Marl		1	0
Red Sandstone Rock	2	2	6
Yellow Sandstone Rock	2	1	6
Red Marl		2	0
Red Sandstone Rock			6
Red Rock Marl		5	0
Red Sandstone Rock		5	0
Red Sandstone & Marl	6	1	0
Red Marl		1	0
Hematite Iron		7	6
Rocky Sandstone, gravelly	1	0	0
Red Sandstone Rock	3	0	6
Light Red Marl	2	2	6
Light Yellow Sandstone	3	1	6
Red Sandstone Rock	3	0	0
Red Marl			6
Red Sandstone Rock	10	1	6
Red Sandstone Rock	2	0	0
Mingled Marl	9	2	6
Blue Rock Binds		5	3
Light Blue Binds	1	0	0
Light Binds	2	0	0
Dark Clod			9
Light Blue Rock		2	3
Brown Binds		2	9
Light Rock Binds	1	0	3
Light Blue Binds	1	2	8
Light Binds	4	1	0
Dark Binds		2	6
Light Red? Binds	3	0	0
Dark Clunch		1	9
Coal			3
Dark Binds		1	6
Light Rock Binds mingled	1	0	6
Light Blue Rock Binds	1	0	6
Light Blue Binds		1	7
Ironstone		2	0
Light Blue Rock		4	3
Blue Binds	1	1	10
Blue Clunch	1	1	3
Coal Smut			2
Clunch with Coal Smut		1	10
Hard Stone Pyrites		1	3
Blue Rock Binds	1	1	5
Blue Binds	2	1	0
Blue Clunch & Coal Smut		2	0
Light Rock Binds	1	2	9
Dark Blue Clunch	3	0	10
Binds	4	2	9
Light Rock	1	2	8
Coal			5
Pricking			2
Light Rock	4	1	5
Blue Binds	5	0	1
Blue Clunch	2	0	0
Coal		6	8
Clod Parting			3
Coal		2	5
Dark Blue Clod [unbottomed]			

Note. Depth to Coal Measures = 136ft, total depth = 343ft. Binds, Clunch and Clod were miners' terms for different forms of clay.

Apart from the plank, the only link with White House Colliery is the trial shaft which may have been the original pit opened up, enlarged and deepened for the purpose. It is near a hedge and full of water almost to the top. The form is oval or more precisely a slightly bulbous rectangle with round ends, measuring 7 × 12ft within. This shape enjoyed considerable popularity in South Wales and elsewhere, as it gave room for both pumping and winding. A rectangular feature shown at one end on the 25 inch map probably denotes an angle-bob pit to house part of the pumping gear, and its alignment with the enginehouse of Newent Colliery may not be without significance.

Flooded and crumbling now, but with 6ft of coal in the bottom, the shaft must have raised high hopes for the new venture. As we shall see presently, the reality proved very different.

White House Colliery. Part of the wooden plank recording the boring, and the oval brick-lined shaft flooded to the top, as it is today.

Aston's Slope

To the east of White House, coal has been worked on a ridge approaching Peter's Farm. Here, a strip of rough ground conceals a small spoil-tip with much blue clay, being the waste from a slope or dipple; there is also a shaft and both are blocked with rubbish. This spot marks one of Aston's trials and coincides with the last remnant of a long abandonded country lane from south of Peter's Farm to Shaw Common. Being on a gradient it had worn down several feet and probably for this reason was never absorbed into the adjoining fields. The site was ideal for a mining trial. From Forster Brown's account the shaft proved 2ft of coal dipping at about 1 in 4 SE, only 12ft from surface. He also reported ironstone 12-15ft above the coal. The dipple was abandoned before reaching the seam. In a field just east of the shaft under clay, a hand-auger recently revealed 4 inches of coal-smut, unbottomed. Thus it seems at least two seams are present, about 8ft apart. The ironstone could not be found; it is probably hematized sandstone traces of which appear in the quarry.

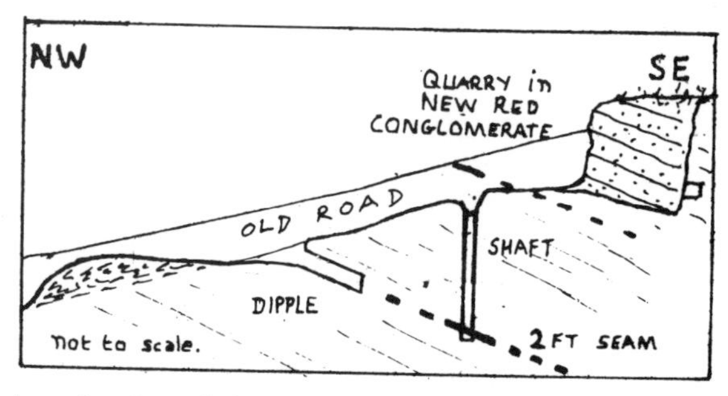

A section through Aston's Slope, a trial working of the 1870s.

Peter's Farm

Peter's Farm is built on Coal Measures and was for long the scene of mining. Two fields to the east were worked, one being named Mine Pit field in 1775. Little is known of these collieries. The area is probably the one which a correspondent to the *Colliery Guardian*[18] referred to in 1879, writing of events at Oxenhall some 10 to 20 years before.

> 'Among the shafts sunk and abandoned was one on Mr. Onslow's estate. A collier employed in the mine was an old Welshman who, when he was discharged, returned to work in a mine a Merthyr Tydfil. What he had seen in the Newent mines clung to his memory and at length he paid them a visit and obtained permission to resume their working. The sinking of two shafts near the site of an old shaft was at once commenced and at a depth of only a few yards, I remember the sinkers came upon a seam of thin coal. Its thickness was irregular, in some places 2 ft and in others dwindling to a mere thread. . . . The coal was like that procured in Dean Forest - oily and flaky, breaking easily, burning well and though giving a good heat, not rapidly consumed.'

The 25 inch Ordnance map surveyed in 1882 shows two shafts in the easternmost of the two fields, one of which was perhaps an inclined shaft described by Forster Brown as follows:

> '*Peter's Slope.* Mr. Onslow opened the dipple here 35 years ago 1847 and worked coal and used it in his own [word illegible] dipple down 27 yds. Aston reopened it, in 33 yds met with three distinct coals pitching steep, bottom coal 2 ft, dirt 4, coal 4 ins. dirt 6 ft other coal 15 inches above. Coal thickened as it went down. Inclination 1 in 3. Company sank a pit here to work coal for engines and used it saw coal on bank. Aston sold 119 tons at Peter's Slope before company took possession.'

It is interesting that a trade directory of 1876 described Aston as Manager of Oxenhall Colliery.

Later Ordnance maps show an additional shaft. One of the three was still open in 1939 and had been sunk by a local well-sinker named Trigg. It was begun in 1893 and finished at a depth of 54yds. Only small quantities of coal were raised by the men employed.[20] This no doubt marked the last serious attempt in the coalfield. Mr Evans of Peter's Farm told me that his father knew a man who had carted coal from one of the pits which was eventually flooded out, tubs and tools being left behind. The circle of a brick-lined shaft still shows in the ploughing and may well correspond to the sinking mentioned by Forster Brown. Near the old railway line a spring emerges from what may well have been an adit or water-level to drain the pits.

Oxenhall Iron Ores and Minerals. *Top*, wood-hematite showing a delicate banded structure; *bottom*, botryoidal and stalagtitic goethite/hematite from underground in the Hillend mine level.

Oxenhall Court - Castle Tump

Coal Measures outcrop in a narrow band running NNE-SSW just east of the old railway bridge, and purple hematite sandstone has been proved below the grass on the hillside above. South of the road in 1775 was a ground called Court Hill, from which a strip was taken by the railway, built in 1883.

According to a One-inch map annotated by Forster Brown, hereabouts was Court Hill Colliery which in 1882 he described as 'old workings of 30 years ago by Mr. Onslow'. It is perhaps the spot which Rev. W. S. Symonds the geologising vicar of Pendock described in 1857.[21] 'A few years ago I carefully examined the pit opened by Mr. Onslow at Oxenhall. The dip of the coal a short distance from the surface was, as far as I now remember, at an angle of 70° . . . I was particularly struck with the coal, some of it being a mere mass of leaves and fibres, very light and not much mineralized.' Strata dipping at 70° was also mentioned by John Phillips.[22] The workings must have been small, for nothing is shown on a large-scale map of 1873.[23] There are several possibilities for the site - the railway line, Court Hill Field, or two encroachments alongside the road east of the bridge. The furthermost is on New Red Sandstone just above the Coal Measures, and a likely place. However, it was occupied by a dwelling in 1841,[24] and later by another, a brick cottage demolished about 1977, which from its appearance dated from 1880-1900. Possibly the workings occupied the spot in the interlude. The only sure conclusion is that Court Hill Colliery has vanished like a dream.

At Holder's Farm the lane is deeply worn down in New Red Sandstone and 200yds west of the buildings a patch of white clay with pieces of coal may indicate a trial. Close to this spot was an old pit 33yds deep and filled up before being cleared out by Aston. He bored down a further 8yds to ironstone but water put a stop to proceedings.

North-east of Pella Farm, John Phillips recorded 'along the line of junction of the Old and New Red deposits red oxide of iron has been dug in a line directed N 35° E Coal was dug here, and its smut is visible with clay at Hill-end.' The latter place consists of two or three houses with a disused blacksmith's shop close to the road. In a field to the south-west a dipple or inclined shaft was at work in the 1860s, and one man is said to have raised over 500 tons of ore in a couple of months which went by canal to the Cwmbran Iron Co.[25] Forster Brown relates that in 1872, from a point 30yds down a level driven 14yds west proved coal 18-20 inches thick but it was soon lost by a fault and never recovered, only ½ ton being

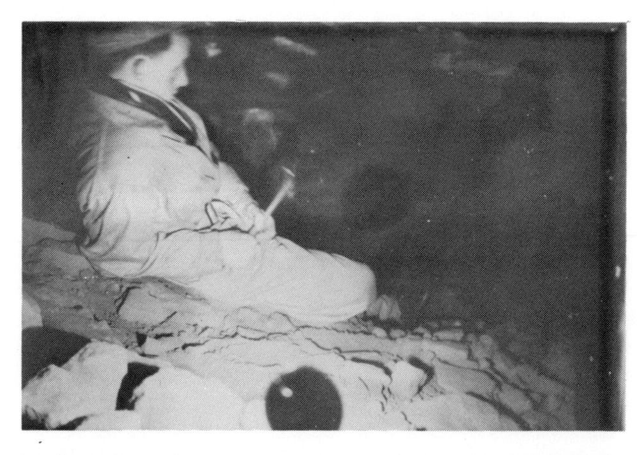

At the bottom of Hillend Mine, 1951, the only known illustration underground in the Newent area. The photograph was taken by the light of a carbide lamp, and the black circle is the silhouette of its reflector.

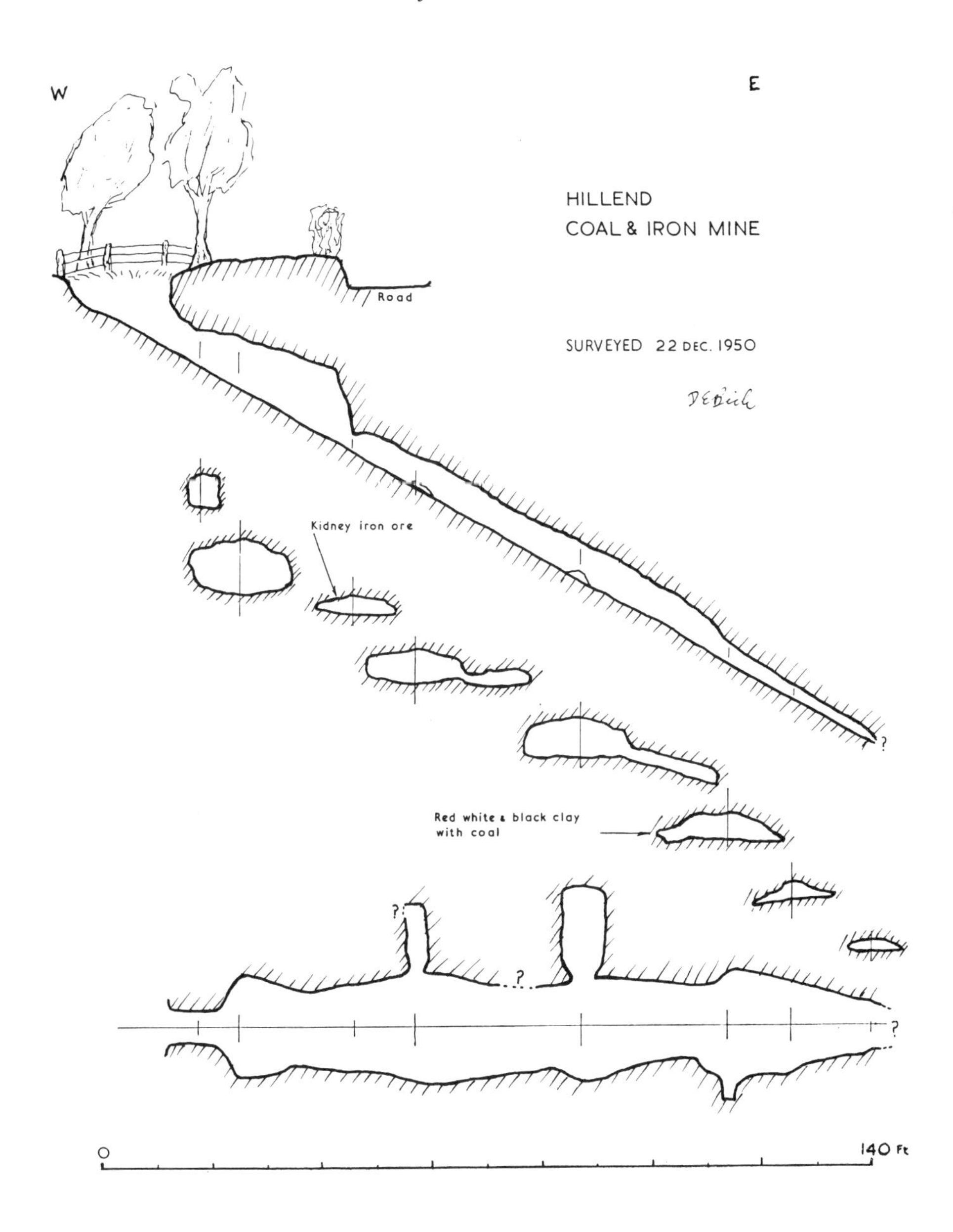

A section of Hillend Coal and Iron Mine, made in 1950. The total extent could not be ascertained due to fallen debris.

raised. On 14 May 1874 the Rev. W. S. Symonds led the Cotteswold Naturalists Field Club on a trip to Newent and when they walked this way from Castle Tump to Hill House Colliery the workings were then abandoned 'the percentage of iron being too small to repay further excavations.'[26] However, under the ownership of J. B. Brown & Co., the *Mineral Statistics* listed it for the same year, and successively until 1879. The dipple was certainly at work in the latter period, for several Forest miners employed there became involved in a misdemeanour at Dymock.[27]

The late Mr Huff of Wisteria Cottage, Hillend, told me that horses pulled the ore out and got so red that they were regularly washed down in a nearby pond, now filled in. According to the late Mr Baldwin of Pool Hill the dipple went in a long way, to water, but at the time of my visit in 1950 so much material had slid down that progress beyond the extent shown by the plan was impossible. Tall trees round an awesome hole marked the entrance to this interesting mine, but regrettably this last accessible working in the coalfield was sealed off some 25 years ago. Its only memorial is purple sandstone in the ploughing, close to an electricity pylon. Similar material also occurs across the road.

The old blacksmith's shop by Wisteria Cottage belonged to Charles Jones who accompanied Forster Brown on his survey. Across the road a depression in the wide verge no doubt originated as an opencast. Aston proved coal here 1ft thick dipping SE, and coal smut has recently been proved on the other side of the hedge. Such a convenient source of fuel may account for the siting of the blacksmith's shop. Behind it, some 80ft from the road, coal was proved 22ft down in a well.[28]

Coal Measures were intersected in driving Oxenhall Tunnel, though with little enough signs of visible coal.[29] Nevertheless, in September 1798 Richard Perkins applied to make a trial in the form of an arch 5ft in diameter. Mineral workings in canal tunnels were then something of a novelty but whether for this reason or because of a general disenchantment, the canal company declined on the grounds of impediment to navigation.[30] In his notes, Murchison recorded that the tunnel was unsupported in the New Red, but arched throughout in the marls and clays of the Old, and presumably in the Coal Measures also.

In 1980 a pipeline trench dug from Newent Pumping Station to Gospel Oak reservoir ran just east of the Coal Measures and proved purple sandstone opposite the turning to Waterdine, and again about 100yds east of Hill View Farm.

4 Newent Colliery

In relatively modern times the only mine in the coalfield worthy of the name was Newent Colliery, though unfortunately liberal supplies of capital, machinery and technical expertise proved of no avail. The early 1870s were prosperous in the trade, but ironically enough a decline had set in even before foundation of the company. With little doubt the catalyst was William Aston who as we have seen, was raising coal on the estate. The money came mainly from wealthy industrialists in Birmingham and Stourbridge, where the father of the Foley empire had commenced in business over two centuries before. It may be that Aston hailed from the same district, though for a time he lived at The Ford, Newent[1] - an address as enigmatic as the man.

By a lease from R. F. Onslow dated 18th December 1875, Aston acquired 'certain mines of coal, iron ore, stone and fire-clay situate in the parishes of Newent, Oxenhall and Pauntley'. The Newent Colliery Co. was formed in May 1876 to take over the lease[2] and to acquire all relevant buildings, plant and machinery. Construction of tramways to and from any railway or canal was specifically authorized, not that the clause was ever invoked. The nominal capital was £7,000 in 700 shares of £10 each, and the firm of William Ridout Wills and Edwin Newey acted as solicitors, with offices at 48 Ann St., Birmingham. Major shareholders were John Betton Taylor who acted as secretary, Joseph John King and Wills himself - a complete list is given shortly.

As we have seen in a previous chapter, borings and a trial shaft were put down on the site of White House Colliery[3] in the summer of 1877, by which time Aston had departed. A small plateau of ground just to the east was subsequently chosen for the new colliery, where two further shafts were sunk. Whether they eventually connected with the trial shaft is not recorded, though it could well have served for ventilation.

In the September, shareholders drew up a new agreement with Onslow to include 925 acres with all mines etc. for a period of 50 years, subject to payment of various rents and royalties. For reasons that are unclear the old company was to be wound up and the assets transferred to a new concern, the Newent Coal and Iron Co., with a capital of £50,000 in £10 shares. Plant included a portable sinking engine and a fixed pumping engine, no coal yet having been reached.

Negotiations were also in progress for a lease of about 100 acres to the south belonging to the trustees of the late Mr Hooke. King acted as agent, arranged the transfer and perhaps settled some debts, for he received 600 paid-up shares for services rendered.

A motive for the Newent Coal and Iron Company's formation was the onset of various difficulties and need for more capital. However, by the middle of April 1878, of the 5,000 shares of £10 each, if we exclude the 600 issued to King, only 476 had actually been taken up, and upon these a call of £4 was made. Thus the real increase in funds only amounted to £1,904 (476 x £4), and most of it came from the original investors. The new subscribers were also men from the Midlands, prominent amongst whom were Frank Evers, Henry Firmstone and Francis Bolton. By this time R. C. Wright had moved to Ledbury and was 'out of business' - presumably retired. He died in 1881. The secretary and colliery manager was a young man named Frederick William Clarke, of Lea Crescent, Birmingham. But within three months an Extraordinary General Meeting passed a special resolution that the nominal capital be further increased to £60,000 by an issue of £10,000 in 10% Preference Shares. This again proved an insufficient inducement and in April 1879 a further issue followed of £3,000 at 20%.

In July 1879 coal was at last struck, one seam being nearly 7ft thick.[4] Great were the jubilations, and the company treated sixty workmen[5] to a dinner at the George Hotel, Newent, with a band from Blakeney playing in the streets.

Unfortunately the seam proved unworkable due to faulting and attempts were made to recover it in more settled ground. By 2nd September only £12,310 had been subscribed and the company was in debt.

The following table shows the shareholders' interests from the beginning. King now had the largest stake, whilst it appears Wright had transferred shares to his son. Wills, who originally appeared as a 'Gentleman' and presumably above trade, later described himself as a solicitor.

Prospects continued gloomy, and in April 1880 an Extraordinary General Meeting held at the Talbot Hotel, Stourbridge, resolved to a nominal capital increase to £73,000 by another £10,000 issue of 20% Preference Shares. How much cash it raised may well be imagined, but worse was to come.

In June the company called in 'a very experienced mining engineer' (probably Forster Brown). His report proved pessimistic. The Newent Coal and Iron Co. was now in a precarious state, not least because it had injudiciously accepted some pretty onerous terms with the lessors. The agreement was as follows, regardless of whether or not the equivalent output of coal had been achieved.

Period	*Royalty Due*
August 1877 - August 1878	£458
August 1878 - August 1879	£608
August 1879 - August 1880	£1,200
	£2,266

One of the two shafts at Newent Colliery, 1965. It is now surrounded in brambles, fenced off and no longer visible.

SHAREHOLDERS IN NEWENT COLLIERY

Name	Address	Occupation	Newent Colliery Co. No. of Shares 5/76	10/77	1/78	Newent Coal & Iron Co. No. of Shares 4/78	9/79
William Aston	Newent	Colliery Owner	10	10			
E. J. Collis	Stourbridge	Mining Engineer	12	12	14	30	30
J. B. Taylor	Clent	Mining Engineer	12	32	65	113	113
J. J. King	Stourbridge	Colliery Owner	12	32	84	216	404
R. C. Wright	Birmingham	Land Agent	12	12	50	120	50
William R. Wills	Birmingham	Gentleman	10	32	84	220	220
Walter R. Wills	Stourbridge	Mining Engineer	2	2	2	4	4
John Colwell	Newent	Contractor			3	5	5
William Rodd	Birmingham	Merchant			10	18	18
G. K. Harrison	Stourbridge	Colliery Owner				100	255
Frank Evers	Stourbridge	Merchant				50	230
H. O. Firmstone	Stourbridge	Ironmaster				50	100
F. S. Bolton	Birmingham	Metal Dealer				100	175
J. Evers-Swindell	Stourbridge	Ironmaster				50	180
W. M. Warden	Birmingham	Merchant					110
R. C. Wright (Jnr)	Birmingham	Clerk					45
			70	132	312	1,076	1,939

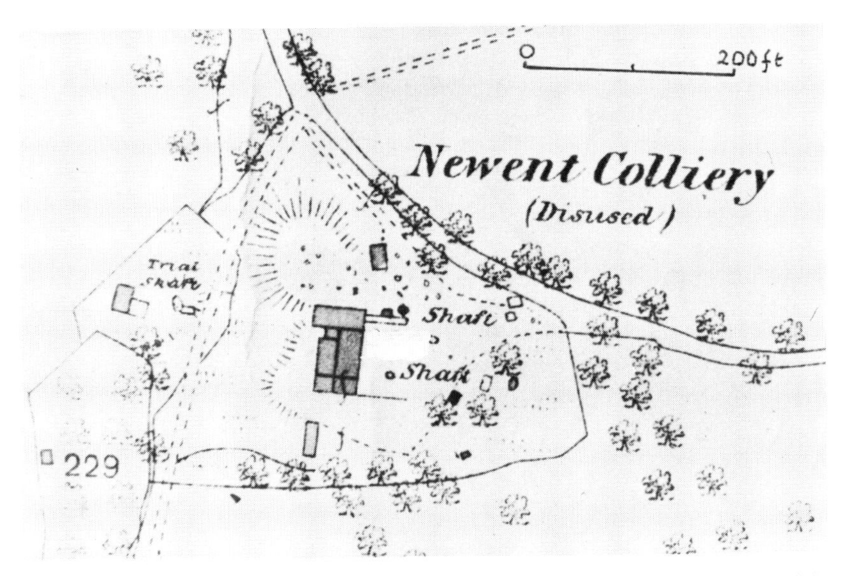

The first edition 25 inch map shows Newent Colliery soon after closure. Note the 'trial shaft' (White House Colliery) on the left.

In fact it appears that precious little coal was sold (the first load went to the George Hotel),[6] and in its predicament the company had few pangs of conscience in dishonouring an agreement which now appeared harsh and unreasonable, especially on an estate where coal had probably never been mined at so much as a penny profit. Negotiations were in hand to ameliorate the terms, but on the death of R. F. Onslow in March 1879 Captain Andrew Onslow[7] refused to pursue the matter until all arrears were paid. This precipitated a crisis, and on 15th July 1880 at the Offices of Messrs Wills & Newey the assembly resolved that 'the Company cannot by any reason of its liabilities continue its business, and that it is advisable to wind up the same.'

No doubt by such an action the company hoped to evade the charges altogether. The action incensed the trustees, who forthwith distrained for the money and proceeded to advertise the plant and machinery for sale on 7th October over the heads of the directors. Clark thereupon filed an injunction and the matter ended in the Court of Chancery. It appeared in evidence that all the subscribed capital, about £17,000, had been spent and that no workable coal existed. However, the company lost the case (reported in *The Times*, edition of 21st October) and the postponed auction by Bruton Knowles & Co. took place on 4th November 1880.

Thus came to a sad conclusion the most determined attempt to work the Newent Coalfield. On this occasion the advocates of perseverance eventually reaping its reward had backed the wrong horse, for even ample coal could not have freed the venture from risk. The coal trade continued depressed and as for transport, with the canal branch long since abandoned, conditions were worse than 70 years before. The Gloucester-Ledbury railway via Oxenhall might have proved invaluable, but was as yet a thing of the future.

In addition, sandstone conglomerates 120-150ft from the surface[7] released vast quantities of water at rates varying from 42,000 to 60,000 gallons per hour.[2] The latter flow is equivalent to 400 kitchen taps, and little imagination is needed to appreciate the problems facing the sinkers. The insurgence was met by the expedient of cutting two chambers on either side of the shaft and installing a steam pumping-engine in one, whilst the other served as a pound or reservoir. The engine chamber witnessed a fatal accident in September 1879 when a rock-fall killed Albert James, a collier from Bream in the Forest of Dean.[3] Although the inflow was thus eventually controlled the burden was a heavy one, with the added charge of an engine labouring endlessly to no profit at all. There was also the cost of deep-pumping and winding, and much of the colliery's output, such as it was, must have gone straight into the boilers.

Mining also introduced other problems. Albert Russell, a miner from Ruardean, found himself on a charge of assaulting Julia Wedley of Kilcot whilst ostensibly looking for lodgings. On being shown the bedroom he made 'insulting overtures' whereupon the lady screamed 'murder' three times. Afterwards he coolly picked a flower from her window-box for his button-hole, but the adventure proved a costly business, amounting to one month's hard labour.[9]

With regard to the extent of the underground workings little is known beyond the fact that 'thin and shattered coals' were followed downwards and eastwards for some distance, until covered by 280ft of water-bearing sandstones.[10] A plan was held by Forster-Brown and Rees, Mining Consultants of Guildhall Chambers, Cardiff, but attempts to trace it have failed.[11]

NEWENT COLLIERY COMPANY.

Within One Mile and a Half of Newent.

Bruton, Knowles, & Co.

WILL SELL BY AUCTION,

(Under a distress for rent)

At the WORKS, on THURSDAY, 4th November, 1880, at Twelve o'clock punctually,—

THE valuable MACHINERY PLANT, TOOLS, &c., including a pair of high-pressure horizontal expansive pumping Engines, cylinders 20 inches diameter, 5ft. stroke by B. Barker and Co., pair Tangye's Patent Winding Engines, cylinders 10 in. diameter, 20 in. stroke, Horizontal Engine for working pump with plummer blocks &c., Savory's improved traction and winding Engine, Cornish Boiler 7ft. diameter, 30ft. 6in. long, with 2 flues, 2ft. 6in. diameter, fitted with Galloway's patent tubes, 3 hemispherical Boilers, respectively 26ft., 27ft., and 35ft. in length, and 5ft. diameter, Boiler fittings, complete Cameron's patent pump, Giffard's patent Injector, massive cast iron bell cranks, pumping spears, Shilton's weighing machine, plummer blocks, new steel wire ropes, Pit Gins, by Bryan Johnson, large quantity flange rails and piping, wrought and cast iron plates, iron buckets, pitch pine timber, &c.

The Engines and Machinery are in first-class working order.

N.B.—The temporary injunction, under which the Sale of this Plant (previously advertised for the 7th of October) was restrained, being now discharged by order of Mr. Justice Field, the above Sale will positively take place.

Catalogues may be had of the Auctioneers, Gloucester.

Money sunk and lost. Glos. Ref. Lib.
The scale of the investment is clear from the extensive plant and machinery.

Strata	147·13 ft above O.D.	Thickness Yds	Ft	Ins
	Raised Ground	1	1	0
	Soil & Sand	2	2	0
	Sandstone	5	0	0
	Clod	2	0	0
	Sandstone	3	0	0
	Clod & Clongomerl	4	0	0
	Sandstone	4	0	0
	Red Marl.			8
	Light Sandstone			
	Red Sandstone	2	1	0
	Marl Partings.			1
	Red Sandstone	2	0	0
	Red Rocky Marl	1	0	6
	Red Sandstone	1	0	6
	Red Sandstone	4	0	0
	Rocky Marl			9
	Red Sandstone	1	0	0
	Dark Pebbly Ground	1	2	0
	Sandstone	2	1	0
	Sandstone Conglomerate	1	0	6
	Rocky Marl		2	0
	Red Marl.		1	6
	Redstone	4	2	6
	Rocky Marl.		1	6
	Strong Dark Sandstone Rook	2	0	4
	Conglomerate	1	1	8
	Red Pebbly Sandstones	6	0	9
	Red Sandstone	1	0	3
	Conglomerate	12	0	0
	Red Sandstone	2	1	6
	Marl & Sandstone	3	1	0

Strata	Yds	Ft	Ins
Sandstone	11	0	3
Sandstone full of Pebbles		2	3
Soft Red Marl		2	9
Light Rocky Marl		2	3
Dark Rocky Marl		2	6
Light Blue Rock			3
Rock Marl	9	1	11
Light Blue Rock			8
Soft Red Marl.	5	0	6
Brown Binds	3	2	0
Dark Grey Binds	4	0	6
Dark Binds	1	0	6
Dark Fire Clay	1	1	6
Dark Blue Binds	1	2	9
Brown Fire Clay	2	1	9
Dark Grey Rock Binds	8	2	5
Iron Stone			2
Dark Grey Binds		1	5
Coal.			1
Dark Grey Binds	4	2	0
Yellow Sandstone	4	1	0
Strong Sandstone		2	0
Light Yellow Sandst	2	1	6
Yellow Binds		1	3
Dark Rock Binds	3	0	0
Binds	2	1	3
COAL	2	0	6
Parting			2
COAL		1	6
(448'-6" TOTAL)	149	1	6

A section of Newent Colliery shaft showing the strata. The left-hand column is Triassic (New Red), descending into the Coal Measures, on the right.

Some reminiscences of the adventure were recorded many years later by Frank Evers' wife Isabel, in the following words:[12]

> 'Another business venture which I remember vividly was Newent Colliery in Gloucestershire. The promoters were chiefly local men, the three Collis men, George Harrison, the three King brothers and Frank, but I think not his brothers. The borings had proved the existence of coal and shafts were sunk and coal of excellent quality found. There was great jubilation and everyone thought his fortune was made. I believe a dinner was held and a huge piece of Newent coal adorned the centre of the table. Alas for premature hopes! After working out a short distance the coal ceased. Another direction was tried with the same result. North, south, east and west were explored, all to no purpose, and at last it became evident that the shafts had been sunk in a little island of coal with a sea of rock round it. What coal there was was worked out and all that could be was saved from the wreck, so that not all the initial outlay was lost, though the loss was severe.'

Frank Evers' son Guy recalled meeting W. F. Clarke about 1923.

> 'He told me that the Newent Colliery affair was one of his first jobs as a young man. The Company worked out and sold the little coal available and then had a stiff fight with the property owners, who demanded royalty as for the whole area, though barren. It seems that the lease was for an area, without mention of whether coal was there or not. John Collis was responsible for this, he believed. Eventually the royalties were compromised for £6,000 and the Company wound up. The assets were not enough to pay all trade creditors in full and it was proposed to go into bankruptcy, but as Mr. Clarke said, "Your father objected strongly to this and urged the others to find the money to pay their debts even if they had lost their own capital. He put his own cheque on the table and eventually the other men followed suit."'

It can be seen from these remarks that the company did not lack a certain dignity, as indeed might be expected when we examine the major shareholders more closely. Most belonged to well known and influential families surrounded by servants and living in baronial halls. Several were JPs, and related by marriage. J. J. King was a partner in King Bros., owners of extensive fire-clay works near Stourbridge, and his wife was a Collis. Investors in the Newent Coal and Iron Co. included King's nephew George King Harrison (1827-1906) who owned several mines and was a friend of John Bright, M.P. He was also a director of the Stourbridge Extension Railway.

Henry Firmstone (1815-1899) inherited the celebrated Leys Iron works at Brockmoor and later owned various other ironworks and collieries.[13] He was also related to the Collis family. Francis Seddon Bolton (1830-1909) of Edgebaston took over his father's copper-smelting works (Thomas Bolton & Co.) which supplied copper wire for the first Atlantic Cable. Frank Evers (1827-1912) was a partner in Cradley Forge and had interests in the Parkhead Furnaces, Dudley, and Horner Hill Colliery. His brother James Evers-Swindell was also involved in local industries and in 1853 was chairman of Bishopstone United Lead mine in the Gower.[14]

In the light of such a background we can see that these men venturing into a forgotten backwater of Gloucestershire were not gullible speculators, but hardened in mining and commerce and with a great range of experience behind them. This being so, it is difficult to see why certain elementary precautions were not taken. Years afterwards, Harrison confided in Daniel Jones a geologist, that 'he had been interested in a sinking near Newent where some £30,000 was lost. After this happened he went to Jermyn St [HQ of the Geological Survey till 1935] and consulted the geologists who showed him the maps of the locality. Had he seen them before he would never have touched it.'[15]

We can only presume that William Aston was a very persuasive character. Possibly the whole of the story may yet survive forgotten in a Stourbridge solicitor's office and waiting to be revealed.

As might be expected, the remnants of the colliery are the most substantial in the coalfield. The grass-grown tip beside the lane to White House is a prominent and not unpicturesque feature which hopefully will be left undisturbed as a monument to the industry. Nearby are two shafts, one fenced and surrounded by brambles, the other and smaller being filled in almost to the top. When ploughed, the vicinity is a mass of broken bricks, clinker and rubble. In 1957 the late Mr Baldwin told me his father helped to build the shafts, the bricks coming from a brickyard about 300yds to the NE and worked by the company; a donkey-engine and tools were left below, and there was a 70ft high chimney. Two traction engines removed the boilers and the weight - or so he claimed - cracked Over Bridge. As to the machinery the newspaper advert is the sole source on the subject, but it well indicates its comprehensive nature. An artist's impression of the colliery in its working days is given on the front cover.

A little to the south where a footpath crosses the Brockmoor brook is a substantial brick aqueduct which carried the canal branch of 1796 over the stream on its way to Hill House Colliery. This interesting relic is worth inspection, not least since a proposed new road from Newent to the M50 at Gorsley threatens its existence.

An unexpected consequence of Newent Colliery was the discovery of large underground springs which proved worth more than minerals. Gloucester had a bad record of insanitary conditions and in 1891 its corporation investigated improved water supplies. In 1893 a borehole was drilled on the advice of Forster Brown SW of Oxenhall Church near the GWR line to a depth of 290ft, being eventually developed as Newent Pumping Station.[16] The Onslow Trustees in a vain hope of finding coal continued drilling to a total of 1,190ft, all in Triassic sandstone. Nevertheless a clause was inserted into the lease, the 'Onslow Agreeement', safeguarding their interests should mining ever again be resumed; a condition which still applies.[17]

5 Dymock

The Parish of Dymock may be said to contain the dregs of the Newent Coalfield, for its measures at surface are scarcely discernible. Nevertheless, coal has been dug in several places.

Castle Tump falls just within the parish, where the main road, sunken after centuries of wear has cut through deposits of coal and iron-ore. On several occasions widening has revealed their outcrop.[1] John Phillips first described the section[2] and Forster Brown wrote as follows:

> 'Coal in well at Mrs Hoghetts about 23ft deep, coal good about one foot dipping S.E. . . . found coal 10 or 12 years ago when sinking lower to find water, worked coal and burnt it, found it very good. Alfred Hickman of Springfield Furnaces Wolverhampton bought land of Hatton, Worcester after coal found, about 3 acres. Red Sandstone Ironstone crops out in road quite near to well showing the Red Sandstone is not far above the coal. . . .'

These observations agree with discoveries made during road-widening in 1977 when 20 or 30ft of bank on the east side was removed to a depth of about 10ft, revealing a great extent of purple hematized sandstone. Into this, two levels or tunnels a few yards apart had been driven from the road-side opposite the gate to Woodbine Cottage. One level connected with a shaft or well (probably Mrs Hoghett's) long since filled in with rubbish, the whole being swept away by the road-works. Some 50yds to the north alongside Castle Tump Cottage I observed the following sections over a length of 20 or 25ft.[3]

Iron ore at Castle Tump, 1977, indicated by dark hematised sandstones. The roof of a level is discernable below the white spot.

Earthy pale red sandstone	4-5ft	Trias
Purple sandstone and conglomerate, charged with iron-ore	1-3ft	
Yellow clay	6ins	Coal Measures
Red and blue-grey mottled clay	2-3ft	
Coal, in large detatched pieces	1ft	

This nearly horizontal section is now hidden behind a high retaining wall. However, purple sandstone and conglomerate are still clearly displayed at its southern end.

The mound constituting Castle Tump is generally explained as a Norman Mott, although the irreverent have ascribed its origins to a medieval coal-tip. Coal occurs between its base and the road, where deep grassy hollows probably denote trials.

Before leaving the area it is worth mentioning that as recently as 1948 a licence to mine here was requested from the National Coal Board.[4] The applicant was the late Charles Marshall of Mitcheldean who had worked at Foxes Bridge, Waterloo and Northern United Collieries in the Forest of Dean, and later in the Kent Coalfield. After the war, with Ronald Meek of the Point Inn (now demolished) he tried to re-open an old dipple east of Nailbridge, called Button-Shank, but found it too dangerous to proceed. Mr Marshall then had various jobs, one leading to the following correspondence, which incidentally reveals the coalfield as a thing unknown to the NCB.

8th March 1948

Dear Sir,

In answer to your letter 23rd Jan. 48. I'm a little late answering, but, I've waited for the Gloucestershire Engineer & Surveyor - Mr. W. E. Evans to draw up a plan of the area where I'm desirous of starting a new mine.

I'm forwarding a map of the area and locality, marked with -X- *OXENHALL,* the second point is marked -X- *Castle Tump.* At Castle Tump this seam crops-out, which would mean a quick producing point, driven into by drift or *slope*, I term it slope.

The reason I desire the two points is that I've actually passed through this outcrop when sub-contracting under a Civil Engineering firm and the coal was dipping between 1 - 3 and 1- 4, nine or twelve inches to the yard also in a state of 7 to 8 ft thick. This is the thickest I've known in outcrop, so it would involve driving direct at this point to Oxenhall, which would save a split in ventilation at shaft workings at *Oxenhall.*

A good point also is that from Castle Tump in slope, intermediate roads can be turned off in a distance of 50 yds: down, 20 or 30 yds: a part driven up hill (pitching) connecting, and on the said dip, *pan faces* could be used, saving machinery, *self-filling*, also in dry condition behind dip (meaning quick production)

Well Sir, I will forward drawing by Mr. W. E. Evans later - (not to hand yet), so I will leave it to your discretion etc.

Yours sincerely,

sgd - C. Marshall.

MEMORANDUM

FROM: Divisional Small Mines Manager. **Our Ref.:** EPJ/BP.
TO: Divisional Estates Manager. **Your Ref.:** K.4/3/G.
Date: 18th March 1948

SUBJECT: application by C. Marshall, High Street, Mitcheldean, Gloc. to open a new mine at Castle Tump, near Oxenham.

Thank you for your memorandum of the 11th March.

I discussed this application with Mr. Prosser, Registrar's Department, in London last week, and I must say he gave me the impression that he was of the opinion that he would be very surprised if coal was found, in the locality in which the applicant intends prospecting.

There is no alternative but to carry out the usual investigation.

I shall now await your news.

For one reason or another Marshall's plans came to nothing, and he last worked in an aircraft factory. He died in 1979 and deserves to go down on record as having made the last attempt (as far as I am aware) to work the Newent Coalfield.[5]

From Castle Tump, blue and yellow clays running parallel to Welsh House Lane denote Coal Measures, but beyond Merehills they contract to mere pockets sandwiched between the Old and New Red formations. Murchison walked over the ground in 1833 or 4 and his observations are worth quoting.

> 'At various intermediate spots along this line, shafts have been sunk and thin portions of coal extracted, but these are entirely wanting in other places . . . To the east of Gamage Hall the carboniferous beds appear for the last time the coal itself cropping out in the ditches of the ploughed lands which lie between the rye-land hills of the New Red Sandstone and the clay of the Old Red. The water passing through the porous overlying strata is held up by the narrow argillaceous zone of coal measures, which is of so cold and heartless a quality as almost to defy improvement. This is the clearest example I am acquainted with of the thinning out of a coal-field. . . .'[6]

One of the shafts mentioned would have corresponded to a spot known in 1814 as Coal Pit Field.[7] It was then the subject of a grant by J. H. Moggridge, late of Boyce Court, and probably marked one of his trials. An Agreement of 1903 identifies the field as lying SE of Little Woodend.[8] According to the Tythe Map it was adjacent to Welsh House Lane, and slight undulations quite near the road probably denote the site, where coal-smut is revealed by a hand-auger. North of Merehills and adjacent to Woodend Coppice the Tythe Map also marks fields called Pit Leasow Orchard and Pit Leasow Meadow, no doubt in allusion to coal workings. Behind Merehills, purple sandstone and conglomerate crop out in a high bank under a hedge, and large pieces of Malvern stone, glacial drift, are also in evidence. Nearby a substantial valley or depression commences, the eastern ridge representing Triassic sandstones.

Our perambulation of the Newent Coalfield comes to an end at Callow Farm, just beyond the River Leadon, where the Coal Measures have thinned out to the extent that Triassic sandstones repose directly on formations of the Old Red.

In the early 1800s the farm belonged to Walter Honeywood Yate of Bromesberrow Place, who consulted William Smith in a fruitless endeavour to find coal. He replied in the summer of 1802 expressing a 'good opinion of your favourable situation.'[9] To turn this to account Yate tried advertising in May 1803 and again in April 1805.[10] And when his Dymock estates were valued in 1807 the forlorn hope still lingered.[11]

> 'An experienced and celebrated Mineralogist Mr Wm Smith of Bath, Engineer to the Duke of Bedford, Lord Somerville etc. has given his opinion decidedly there is a vein of coal on the Callow Farm and it has also been bored and sufficient Proofs appeared to confirm this opinion. Its distance from any other collieries and the contiguity of great towns and water conveyance . . . renders this a most desirable situation for a Colliery and consequently must much increase the value of the Property.'

Whether coal was ever actually proved is doubtful, indeed there was no specific statement to the effect. The estate passed into the hands of the Beachamps, and after a chequered career poor Yate ended in an asylum.

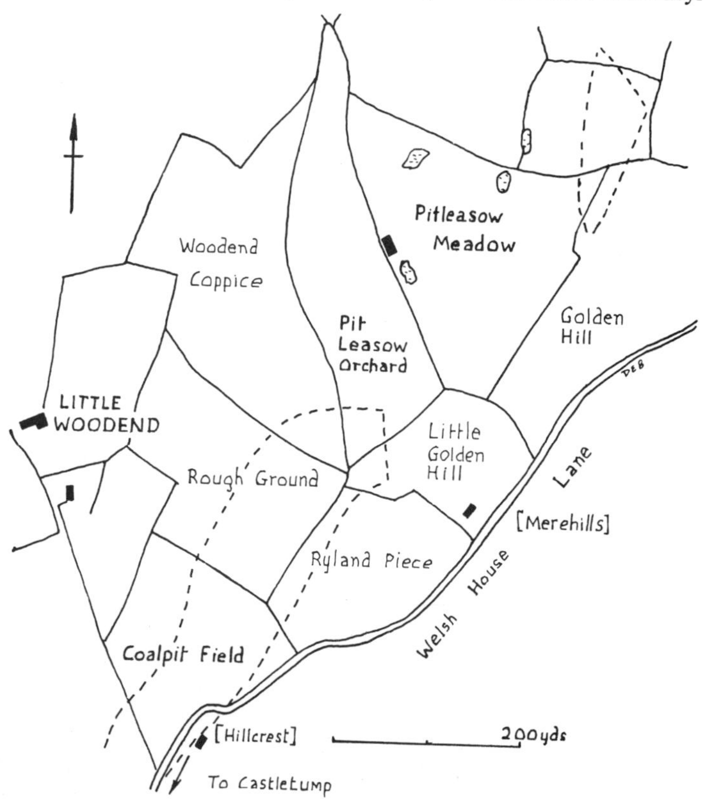

The Dymock tithe map of 1843, with field names. The broken line denotes the outcrop of the Coal Measures.

Part 2 MINING OUTSIDE THE COALFIELD

1 Trials for Coal

The record of coal mining near Newent would be incomplete without mention of trials in areas where it is now known that none could ever be found. But in days when geology was little understood, the search depended less on science than wily mine-agents and promoters adept at persuading landowners of their chances. Few indeed were the parishes that had not witnessed serious attempts or at least prospecting at one time or another.

Dealing with such doomed adventures we will begin a few miles from Newent where the Malverns rise up behind Bromesberrow, at a rocky wooded knoll called Coal Hill. It lies just west ot Chase End Hill, and probably belonged to W. H. Yate of Bromsberrow Place, to whom the trials might be attributable. The name is a legacy of shafts sunk in shales of the Cambrian age, which by their black appearance tempted the prospector. 'The dark colour of this shale', wrote John Phillips, 'must be the excuse for several vain trials for coal made in the low ground about Whiteleaved Oak. Some of the pits may still be seen.'[1]

Recently, in directing me to Coal Hill, a local man whose ancestors had lived in the district for generations volunteered the information that according to an old lady now deceased, coal could be found there for the taking; he saw no reason to doubt it. Such is the eye of faith, and such are the tales that linger in this beautiful and sequestered corner of Gloucestershire.

At Upleadon in the heart of Triassic marls is a very early reference, dating from the reign of Henry II (1154-1189). Here, at Eadulverfshelle (Edenshill) the monks of Newent had the right to 'coal in that wood, to plough it up, if it be not within our forest.'[2] Curiously enough a deed of 1778 concerning the Manor of Carswalls, close by, mentions three grounds called Colepitt Fields and Colepit Meadow.[3] These no doubt conform to Great and Little Coalpit pieces, and Coal Pitts between Edenshill and Little Carswalls, so called in 1840. Today these fields contain several big hollows, often water-filled, and presumably abandoned marl-pits. One however, in a ground now known as Coalpit Field (Kittlebury in 1840)[4] is about 10ft deep and 80yds long, dry in all weathers and according to tradition, a coal-working. Since ridge and furrow ploughing descends part-way down the sides it must be of some antiquity. On investigation, a hand-auger revealed 4ins of black smut about 2ft below the bottom. But careful inspection suggested it was the residue of charcoal-burning. This occupation was common during the time of Newent Ironworks, and probably accounts for the field names. But it does not explain Edenshill.

South of Newent, in October 1867 the *Mining Journal* reported that

> 'A shaft is being sunk at the base of May Hill, 8 miles from Gloucester. It is said that some 60 years ago a pit was sunk near the present shaft to a depth of over 30 ft by working men who had great difficulty out of their limited means to accomplish what they did. One night a party of Foresters it was thought, came and nearly filled up the shaft. The men were so disheartened they did not renew their attempt.
>
> There are several old people who persist in the accuracy of the foregoing and further add they carried away coal from the old pit, and this led a person named Barnes to offer a piece of land to a party acquainted with coal matters to make another experiment. On Tuesday the sinkers got down 36 ft without any traces of coal. . . .'

A fortnight later the shaft was said to be 50ft deep 'with no indication of coal or any gaseous substance.' The principal promoter was Mr Benjamin Stephens, an old collier. The site of these trials is now lost.[5]

Apart from a little coal at Howle Hill, Herefordshire is too old geologically for such deposits. However, before such matters were understood, in 1809 a meeting was held in Hereford to subscribe for funds for prospecting, and some work ensued including trials at Checkley.[6] However the possibility of small pockets of Coal Measures in the Old Red formations should not be altogether discounted, and may explain an unexpected discovery at Gorsley. A contractor told me that when digging a hole to bury tree-stumps some years ago at Beavan's Hill (676249), he came across a seam of coal 4-6 inches thick and about 7 feet down, beneath a bed of brown and yellow clay.

2 Iron Mining and Ironworks

'I confess that these old metallurgical processes have a peculiar charm for me . . . something which has done good service in its day has gone forever.

John Percy

The iron industry of Newent and district falls into three distinct periods:

1. Quarrying or mining ores and smelting in bloomery hearths during the Middle Ages and very much earlier.
2. Elmbridge Furnace, Newent. Re-smelting the rich slags or cinders, together with local ores (c1639-1751).
3. 19th-Century mining for smelting outside the district as described in the Oxenhall chapter.

Little is known of the first period, but with ample woodland to supply charcoal for the leisurely rate of production, nearby sources of ore must have been the main factor in choosing a bloomery site.

Over the centuries the cinder-heaps grew to huge proportions, and the quantity of ore smelted may be inferred from a Worcestershire man, Andrew Yarranton, writing in 1677 'In the Forest of Dean and thereabouts as high as Worcester, are great and infinite quantities of cinders . . . which will supply the ironworks some hundreds of years, and these cinders make the prime and best iron, and with much less charcoal than doth the ironstone.'[1]

From the records it would be possible, though a task of considerable labour, to ascertain with more or less accuracy the quantity of cinders within a few miles of the town, and present indications suggest 50,000-100,000 tons. The equivalent ore would have been greater still, though whether derived from the Triassic sandstones or from older rocks near May Hill is impossible to say, except perhaps, by analysis.

A century after the slags had been removed, their memory lingered on in field names, as below: [2]

Field-name	*Locality*	*Parish*	*Grid reference*
Cinder Pits	Botloes Green	Newent	720283
Cinder Meadow	Nelfields	Newent	730246
Cinder Field	Nelfields	Newent	729249
Middle Cinders	Haynes Farm	Taynton	745222
Lower Cinders	Haynes Farm	Taynton	746221

Bloomery slag is still often revealed in ploughing and has a smooth dull, dark gray surface. It is noticeably heavy due to the iron content and often displays blow-holes.

The second phase of Newent's iron industry lasted well over a century. It was the presence of cinders which enabled a nationally important era of iron-working to become established in and around the Forest of Dean, the Forest aspects of which have been described elsewhere.[3] For generations these ironworks played an important role in the economy of West Gloucestershire. They gave employment to large numbers of wood-cutters, charcoal burners, miners, carriers and boat-men on the Severn, not to mention those at the furnaces. Since one ton of iron devoured some 10 tons of wood, the labour on this side of the operation was very great. Charcoal for Elmbridge came from as far afield as Ashperton, Woolhope, Putley, Eastnor and Longdon. The demand provided a useful bonus for farmers and landowners; it encouraged proper management on a rotational basis, and contrary to general belief did not destroy woodland any more than harvesting destroys a cornfield.

Acting as a flux, the cinders with the added bonus of their iron content, rendered the recalcitrant Forest ores more fusible.[4] The latter were not however, much used at Elmbridge in spite of a general belief to the contrary. Besides Elmbridge Furnace, others had sprung up at Longhope, Ross, Whitchurch and St Weonard's, but in contrast to those in the Forest of Dean, their history is in general obscure.[5] Elmbridge and St Weonard's became absorbed into the Foley complex and most of the pig-iron went overland to the Severn for conveying upstream to various forges for conversion to malleable forms. Newent iron was carried via Upleadon to the banks of the river at Ashleworth, where storehouses were established for its reception. Some of the iron was converted in a forge on the River Leadon, more latterly a water-mill, and the name of Forge Lane, Upleadon, is a legacy of this period. Upleadon forge was held by Humfry Soley in 1679/8, and by John Soley from 1703/4. George Draper was in occupation after 1715.[6]

Of all the Gloucestershire and Herefordshire furnaces, Elmbridge was unique in the exceedingly low proportions of ore to cinders in the charge, and its reliance on local (non-Forest) ores.

An invaluable source on these ironworks are the Foley papers, some 2,000 cardboard boxes of which are deposited in the Record Office at Hereford.[7] The furnace accounts reveal a meticulous attention to detail, and apart from ten years are virtually complete for Elmbridge from 1692 to the end of the Foley involvement, which almost certainly heralded the closure of the works.

An idea of running costs can be gained from the following abstract for a typical year, and the overwhelming burden of the charcoal fuel will be noticed:

Elmbridge Furnace Costs, Sept. 1693 - Sept. 1694

	£
Charcoal	1,966
Myne (ore)	43
Cinders, including washing	424
Wages and salaries	201
Maintenance of furnace, etc	131
Rents and sundry charges	195
	£2,960

Iron production, 542 tons. Cost per ton =£5.46

As to the ore, we find that over the whole period it originated largely from 'Mr ffoley's land at Aston [Ingham]', and there is no reason to suspect this forgotten source was not worked from the beginning, in the reign of Charles Ist nearly 50 years before.[8]

Interpreting the accounts is complicated because the ore, cinders and charcoals were measured by volume, not weight, the unit being the Dozen Bushel, or Load. All that can be said at present is that from what we know of the furnace output and the probable content of the cinders, the Load must have been about a ton.[9]

The graph shows the furnace performance in terms of cinders, ore, and iron output. Charcoal consumption was generally on a par with cinders, at least between 1692 and 1716.

The second graph is in some respects more interesting. The upper line shows the iron/charcoal ratio, which may be called the furnace efficiency factor, and in view of thc high cost of charcoal the later years must have given cause for concern. The lower line indicates the ore/cinders ratio. This fell to an astonishingly low value of 0.12 in 1701-2 and only averaged 0.24 throughout. With such proportions we may well ask, which was the flux and which the ore! By way of comparison, at this period the Redbrook and Bishopswood Furnaces under the same management as Elmbridge ran at a slightly better efficiency factor on an ore/cinder ratio of about 0.40. It might be instructive to establish whether the latter ratio for the various furnaces depended in any way on the source of the cinders.[10] This is an avenue I have not had time to explore. A full statistical analysis might yield very informative results, but the task is complicated by many factors.

It must never be forgotten that in spite of all the skill and experience which could be mustered, smelting operations of this kind were renowned for their unpredictability. As one manager expressed it, 'A furnace is a fickle mistress and must be humoured and her favours not to be depended upon . . . the excellency of a Founder is to humour her dispositions, but never to force her inclinations.'[11]

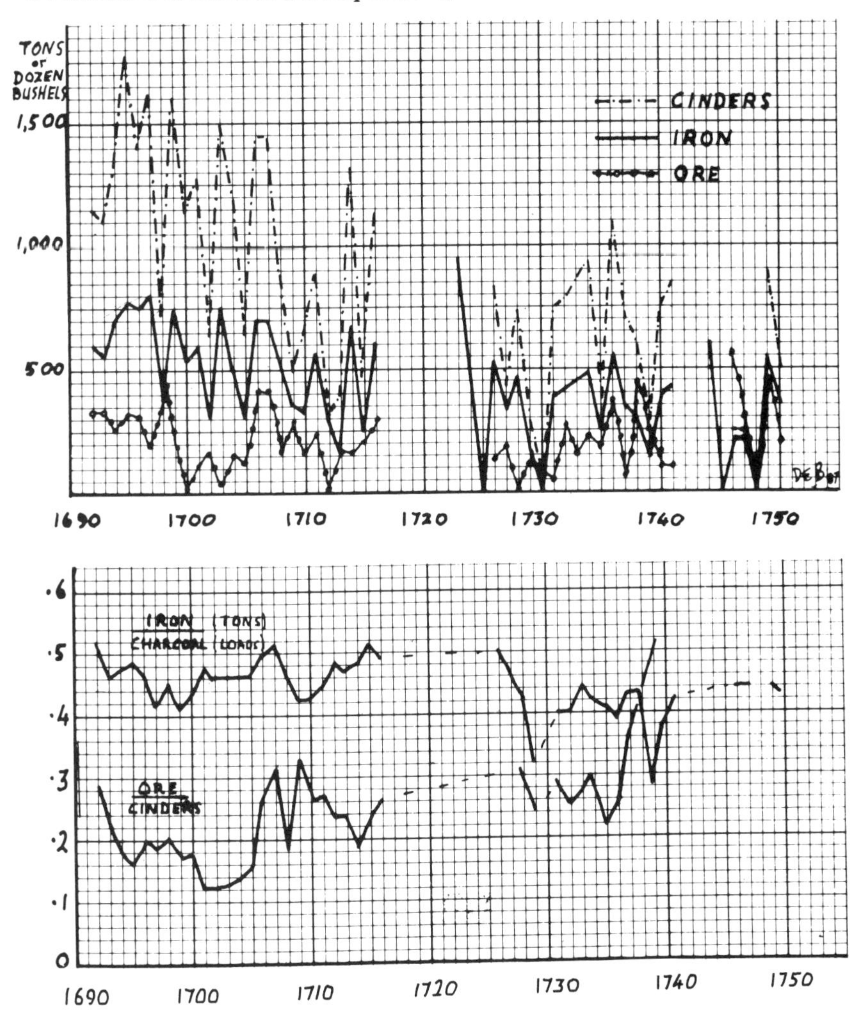

Statistics of Elmbridge Furnace, as derived from the Foley archives.

The remnants of Elmbridge (Newent) Ironworks, showing the blowing-house, in front of which stood the furnace. The charcoal-store is in the background.

To supplement Aston ore, though rarely to replace it altogether, other sources sometimes appear in the records. The reasons were more probably metallurgical than economic, and the idea that Lancashire hematite was only brought in, and at great expense, due to the exhaustion of Forest ores, seems rather wide of the mark.

Non-Aston imports to Elmbridge are given below.

Year	*Source*	*Dozen Bushels*	*Year*	*Source*	*Dozen Bushels*
1692	Dean	210	1738	Dean	206 tons
1693	Dean	10	1739	Dean	19 tons
1693	Unspecified	33	1746	Dean	579
1706	Dean	300	1746	Black House	294
1707	Dean	304	1749	Dean	435
1716	Newent	50	1749	Black House	41
1726	Lancashire	93 tons	1750	Dean	159
1727	Lancashire	294 tons	1750	Black House	64
1727	Unspecified	166			
1729	Unspecified	140			

Opposite: A page from the Foley Partnership accounts for the year ending 29th September 1730, relating to Elmbridge Furnace.

From ye 29th Sept 1729 To the 29th Sept 1730 — Elmbridge Furnace Acct.

		T.	C.	qr.	lb.	£	s.	d.
Sow Iron	In Stock 29th Sept 1729	221	6	1	14			
	Made in the Time of this Acct.	119	2	2	21			
		340	9		7			
	Whereof Deliver'd							
	To Mr Jno Baxter pigs at 7£ & Castings at 8£ per Ton	13		2		94	4	
	To Mr George Draper at 6:14 & 7£ per Ton	30	5		21	206		
	To Mr Tho: Harvey at 7£ per Ton	20				140		
	To Mr Edw: Knight at 7 & 6:17 per Ton	45				312	15	
	To Mr Nath: Mason at 7 & 6:17:6 per Ton	75				521	17	6
	To Lidbrooke Forges	1				1274	16	6
	To Retail Castings Sold by Jno Jones		2		11			
	Remd Old Castings 7:14:1:14 New 1:3:17 pigs 148:5	156	1	1	3			
		340	9		7			

		Coles Lo.	Sa.	Brais Lo.	Sa.
Charcoles	In Stock 29th Sept 1729	253	5	185	11
	Recd from Several Woods	414	3	34	4
		667	8	220	3
	Whereof Spent to make 119:2:2:21 of Sow Iron	366	4		
	Brays sold in the yard per Jno Jones			5	8
	Remaining	307	4	214	7
		667	8	220	3

		Dozn.	B.
Cinders	In Stock 29th Sept 1729	332	5
	Recd from Several places	1185	4
		1517	9
	Whereof Spent to make 119:2:2:21 of Sow Iron	288	6
	Remaining	1229	3
		1517	9

		Dozn.	B.
Mine	In Stock 29th Sept 1729	11	10
	Recd by Several persons	140	4
		152	2
	Whereof Spent to make 119:2:2:21 of Sow Iron	69	5
	Remaining	82	9
		152	2

For many years following 1724 Aston ore was supplied by either Thomas Warr or his widow ('Widow Warr'), both of whom were perhaps tenants of Lord Foley. The 1716 source is listed as 'myne from Mr ffoley's land in Newent'. This may have been a reference to Black House near Glasshouse, which is specified in 1746-1750. Black House was then also providing 'flux mine', probably limestone to assist smelting the Dean ores, and implying the Aston workings had become impoverished.

Due to muddled accounting over the last few years, the figures are difficult to fathom, a problem which interestingly enough was encountered at the time, when the manager Robert Leech found himself in trouble with Thomas Pendrill, the steward to Lord Foley. From this latter period an example of the cost of ore is given below. (The price of Aston ore remained at 3/- or 3/6d for many years.)

	Doz	Bushels			*Cash paid for Mine (ore)* £	s	d
For	186	7	of fforest mine	at 10/- per Doz	93	5	10
	215	4	of Mine	at 3/6 per Doz	37	13	8
	78	10	Ditto	at 8d per Doz	2	12	6
	Jos. Grindall & Brother for getting Mine which is in stock at the Black House				6	14	0
					140	6	0

Whether at 3/6d or 10/-, the cost relative to overall charges was small, so that if Forest ore had offered any real advantage it would have been used on a regular basis. We may conclude that trials in this direction were a failure, for they were not continued, except in a rather random fashion. It is of course possible that their particular characteristic was occasionally preferred. Altogether we are left with a sense of mystery regarding the metallurgy of Elmbridge Furnace - an enigma which Professor B. L. C. Johnson noted over thirty years ago.[12]

In summary, for the fully documented periods the total figures are as follows. The Aston total is conservative since some of the output was almost certainly included in the unspecified sources.

Elmbridge Furnace Statistics 1692-1716, 1728-41, 1745-51

Ore from Aston	6,900	Dozen Bushels (Loads)
Other ores, counting a ton as a dozen bushels	3,070	Dozen Bushels
Total ores	9,970	Dozen Bushels
Cinders	37,800	Dozen Bushels
Charcoal	43,500	Loads
Iron production	19,353	Tons

Average ratio, ore/cinders = 0.26, iron/charcoal = 0.44

Average yield of iron from ore and cinders together = 40.5%, assuming a dozen bushels to the ton. This recovery is very similar to that obtained elsewhere from Forest cinders.[13]

If Aston provided ore at the same rate throughout the whole life of the furnace, the total output may be computed at some 21,000 Dozen Bushels, or about the same number of tons. It is thus no wonder that signs of such substantial operations are still to be found as we shall see shortly.

As for the reasons for final closure, it is interesting to observe that in the last years, when Aston ore was no longer used there is evidence of more experimenting, as if technical difficulties had arisen which could not be resolved. A dearth of charcoal was hardly a problem, for on a properly managed basis supplies could have held indefinitely. More probably a growing shortage of cinders was another factor, for eventually supplies had to be carted from the Forest, and 16 or 17 miles from Whitchurch. Even so, reserves were never exhausted, for thousands of tons still exist over a large acreage immediately south of Peterstow near Ross, and elsewhere.[14] The last charcoal furnaces in Gloucestershire did not in fact close until the beginning of the 19th century.

It is an interesting exercise to trace the remains and physical evidence of the Newent and Aston workings, of which a number of smaller excavations and trials are no doubt lost forever.[15] The Foley accounts are reticent regarding exact locations or methods of extraction, but fortunately we have other clues. The Black House site almost to a certainty corresponds to fields north of the farm where 'Old Quarries' feature on Ordnance maps, and which in 1774 were called Ore Piece and Ore Close.[16] The spot was visited on 27th December 1834 by Stephen Ballard, the engineer to the Hereford & Gloucester Canal Co, and a very reliable witness. His diaries reveal important evidence:[17]

> 'Went with Mr Ellis and Mr Lewis to see the old Iron Stone mines at Black House and at the Ore Piece about 2½ or 3 miles from Newent, the one at the ore Piece belonging to Mr Ellis. I should think it is very rich. The rock in which it crops up near the surface and the Ore was obtained by open work, quarry fashion, here and at Black House. Also saw Crokes hole at Cugley, a subterraneous passage cut in the sand rock. It is not known when this passage was cut nor for what purpose. It extends under the earth a great way. Old Mr Collar of Wood Gate says he remembers that some miners who were employed at the coal works at Boulsden went in a great distance but not to the end, the damp having prevented their progress. . . .
>
> I think it is not improbable that this was an entrance to an Iron mine, the passage is in the direction of the Black House, where iron stone was procured for the Furnace at Newent . . .'

The workings appear to have been in Wenlock Limestone which also would have supplied the flux myne, and it is a pity they are now filled in, rendering examination impossible.

The tunnel at Cugley was still open within living memory, but the entrance is now marked only by a hollow. The location is confirmed by Bryant's county map of 1824 and early 25 inch Ordnance maps. Apart from mining associations it is of historic interest, being the Crocket's Hole mentioned by Rudder where one Crocket hid in the persecuting reign of Queen Mary.

Although the name Ore Piece occurs in Ballard's account, the context imputes it was not the Ore Piece at Black House, but somewhere else, 2½-3 miles from Newent. We can scarcely doubt that the place is north of Oaks Lane between Aston Ingham and Clifford's Mesne, where Nine Acres Wood is marked on the

1840 Tithe map. Next to it is Ore Field, belonging to John Ellis who was presumably Ballard's companion on his tour of old mining sites a few Christmases before, and the same John Ellis who ran The George Inn, Newent.

As at Black House, Wenlock Limestone occurs both at Ore Field and Nine Acres Wood; its surface is still scarred with deep grooves and pits akin to run-in mine shafts complete with venerable yews, bearing a strong resemblance to Roman iron-workings or Scowls in the Forest of Dean. The trenches must represent open-casts on the outcrop of iron lodes. The longest workings end at a footpath in the wood where the Old Series geological map denotes a major east-west fault. The hollows in Ore Field are gradually being filled in; the east bank is still 15 or 20ft high in places, and hematised limestone can be picked up in the ploughing where a public path runs in the next field. Genuine ore samples would be a valuable find, for apart from our present interest, minerals in economic quantity are rare in the Wenlock of this country.

This neglected site at Aston, not to mention others perhaps awaiting discovery, is also important to industrial archaeologists, not least being the question (in spite of Ballard's evidence) whether the deeper workings resorted to underground mining. Another interesting question arises from an entry in the Foley

'Mr Foley's land at Aston', as it is today, showing the trench-like workings.

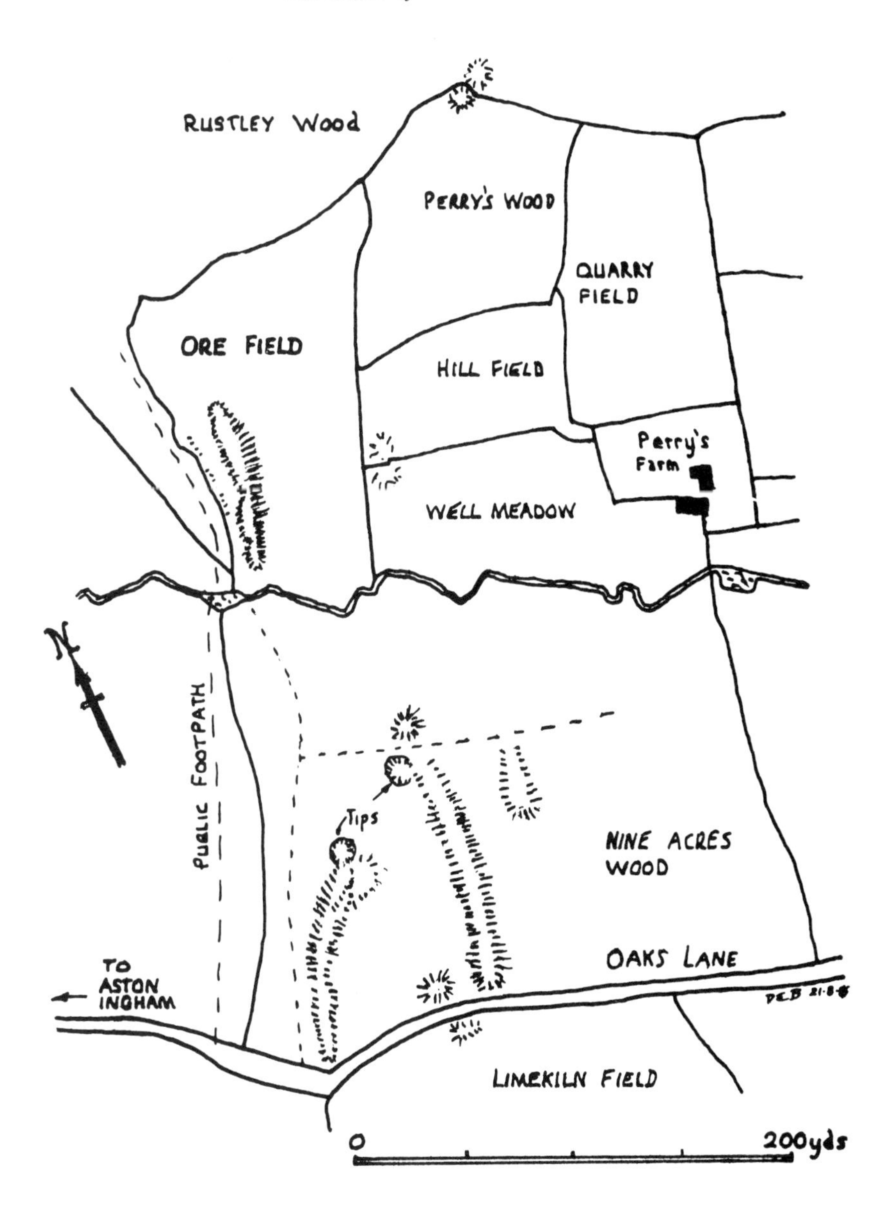

Ancient iron-workings. HRO
Based on the tithe map, this shows the present evidence of long abandoned excavations in wood and scrub-land east of Aston Ingham.

accounts dated November 20th, 1746. It concerns a payment of 13/8d to James Marshall 'for making a Gate and a Stile on account of the Carriages to carry mine from Aston to Bishopswood.' Again we must wonder, what were the magical qualities of Aston ores to justify carting many miles over rough and hilly roads to a furnace on the banks of the Wye. And we may also wonder whether Marshall's gate and stile, or rather their replacements, still survive.

At any rate, these wooded acres just over the border into Herefordshire were the scene of constant activity and employment in the reign of Charles I and for generations afterwards. They deserve a mention in the annals of history.

Although somewhat beyond the scope of this volume, something will be added on Linton Furnace, already mentioned. The Parish Register refers to a 'ffurnisse' as early as 1630, and it doubtless utilised slag from the vast Roman smelting site which covered 250 acres at Ariconium (Bromash) close by.[18] The furnace came into the Foley empire in 1682. It was acquired by Paul Foley for £700 from Robert Chelsham of Old Swinford on behalf of Henry Cornish of London, on land called Ladycroft together with the 'Buildings and Houses thereon erected.' The sale included rights to a 'ffurnace poole' and a watercourse through the lands of John Sargeant.[19] Linton Furnace yielded 543 tons of iron in 1686/7 but had gone out of blast a few years later.

The site of these works has long aroused speculation.[20] To power the water-driven bellows a situation by the Rudhall Brook somewhere in the parish between Aston Mills and Harleton Farm is almost certain. However, unlike the long lost Ross Ironworks (c 1588-1659)[21] we are faced with not too few possibilities but too many. Indeed, in this 'valley of mystery' we are driven to suspect more than one furnace, although perhaps active at different periods.

The evidence may be presented by starting at Aston Crews and working downstream. West of an ancient road leading to Burton Court, called Cut-throat Lane, the Tithe map discloses grounds named Near, Lower and Far Furnace Hill. Lower down, the road crosses the Rudhall valley by an embankment, the north side of which is retained by a massive wall. The wall has features suggesting another purpose originally, perhaps part of an ironworks. The embankment may have been built as a dam, the road making use of it. There are in fact signs in the form of a distinct shelf that a road or leat once continued along the east side of the valley.

Just above where the road crossed the stream are clear remains of a watercourse or mill-race, the upper reaches of which have been bulldozed away. Also, in the bank of a nearby cart-track is a long wall of red sandstone blocks. The field next to the embankment reveals much grey-green charcoal furnace slag when ploughed, and bloomery slag occurs in the ground above, adjoining Cut-throat Lane.

In 1936 Rhys Jenkins proposed this vicinity as a site for Linton Furnace, and not without reason.[22] There is however a better candidate, as we will see shortly.

Below the embankment is a marshy ground and a further embankment with remnants of a sluice, though for what purpose is not certain. A little down the valley is Burton Court where intriguing evidence has recently come to light. In 1719 a field nearby was known as Furnace Close,[23] and the Parish Register for

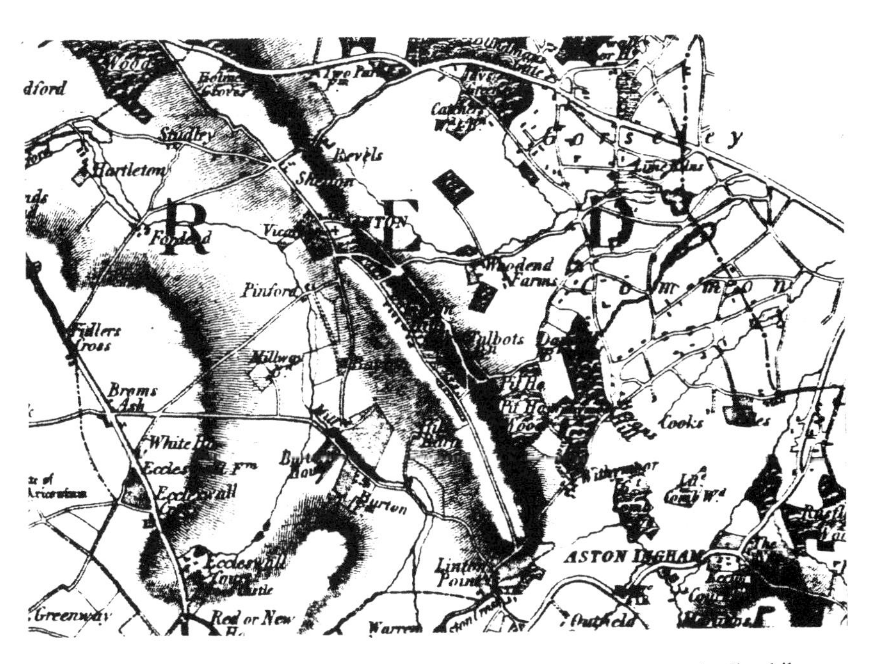

The Rudhall Valley - a detail from Bryant's map of 1835. Note the limekilns on Gorsley Common.

1688 mentions 'Burton's Furnace'. The oldest building here may be termed a barn, for want of a better name. It is very long, with several bays or divisions, and considerably below ground level at the upper end, where there is a large well. An origin based on agricultural usage seems quite improbable. The barn lies at more or less right angles to the house and a few yards from it, though the gap has long ago been occupied by an extension. It clearly dates from the 17th century or earlier, and with various doorways and windows either inserted or stopped up, testifies to several changes of use. Apparently its attics were at one time living accommodation. On the side next to the house the ground level is much higher, having been built up, perhaps from the hardcore of demolished buildings, to form the drive.

Another enigmatic feature is a tunnel about 4ft high and with an equal cover, running parallel to the barn and under the extension, which it undoubtedly pre-dates. The passage is cut out of solid rock, the pick-marks being abundant. It penetrates under the dwelling 17ft and inclining to the left, where the width has been reduced and the passage finally stopped up by walling either in brick or stone, perhaps for structural reasons. At the inner end a blank face of sandstone is exposed, where it appears the tunnel finally turned sharp left,[24] bringing it very close, if not actually into the barn, the floor of which is a little higher. In the other direction, the tunnel turns abruptly north-west, but a fall of earth at the corner prevents further exploration. From this point it was probably constructed on the 'cut and cover' principle. On the same alignment and on a lower level about 200 yards distant, a stone-lined tunnel or culvert passes under the road to Bromsash and discharges into the Rudhall Brook alongside a small shed. There is little doubt that the two tunnels linked up, for the one beneath the road has a branch heading straight for Burton Court; it is now to a great extent silted up.

Hereabouts on both sides of the road the fields display much slag when ploughed, bloomery slag on the south and charcoal furnace slag on the north. In the former area pieces of furnace-lining have been found.[25]

If the tunnel does indeed connect with the culvert it must have been a watercourse or tailrace, and bearing in mind the peculiarities of the barn, we are left with the strong suspicion that these mutilated features are the remnants of the Linton Furnace complex with its buildings and houses as mentioned in 1682. As to the exact site of the furnace itself, speculation is idle without more research; at all events, it is my belief that these few acres hold the key to the puzzle.

One problem to be resolved concerns the watercourse for working the bellows, and here it must be admitted that difficulties arise. The Foley accounts for 1686/7 refer to watercourse rents amounting to some £34, payable to John Sargeant and five others.[25] This implies a considerable length and is not altogether commen-

The 'barn' at Burton Court. It probably served some purpose for Linton Ironworks - perhaps a charcoal store and offices.

surate with Burton's situation in the valley. By way of comparison, the rent for the two-mile leat for Newent Ironworks was only £17-£20. On this basis a site further down the Rudhall Brook would be expected, and there is indeed some evidence in the form of a field name.

Just north of Harleton Farm the Tithe map refers to a ground called Furnace Piece, but unfortunately it has been covered to a great extent by the Ross Spur. On the debit side, no appreciable slag has been found, nor signs of a watercourse.[27] Thus we are again confronted with an enigma; like all the other furnace fields, the name may be a red herring or it may not.

After the passage of three centuries and thirteen monarchs, we must not be surprised if the Rudhall Valley remains forever beyond our understanding.

3 Trials for Silver and Gold

Finally, we pursue the more exotic subject of mining for precious metals. Serious attempts were made in the 1670s and 80s, but once again it is clear that the records which survive are but the tip of an iceberg.[1]

The long-standing tradition that gold was worked in the parish of Taynton, south of Newent, is confirmed as follows:

> 'About 1680 a gold mine was discovered at Little Taynton in Gloucestershire. The Society of Mines Royal seized it and granted leases to refiners who extracted gold, but they did not go on, as the gold sometimes would not repay the charge of separation though often it did.'[2]

A century later Rudder made local enquiries but without result.

Rocks Wood is a possible location for these operations, which probably worked alluvial deposits. This wood occupies a narrow and deep valley worn down by the Glasshouse Brook which divides the parishes of Newent and Taynton. An ancient leat or watercourse contours its southern slopes and may have had associations with the works.

Recently the stream has been carefully panned by an expert in this difficult art, and minute traces of gold have indeed been indicated. It would be instructive systematically to examine the area, for gold certainly occurs (though not in economic amounts) in the Forest of Dean.[3] (See also the next chapter).

Just above Rocks Wood the Glasshouse Brook cuts through Wenlock rocks and is joined by another stream. The ground between the two is waste land on the edge of Taynton parish and much overgrown, and by local repute once a lead mine. The excavations form a long hollow in limestone which outcrops 8-10ft thick in the Glasshouse Brook, forming a waterfall. From here to the intersection of the streams where there is evidence of a shaft, mudstones, knobbly limestones and clays lying nearly horizontal are exposed on the steep bank. There are similar outcrops just below the confluence and on the side of a cart-track a few yards to the north.

In these exposures galena (lead ore) and blende (zinc ore) are scattered in small amounts up to the size of a match-head. These materials are also found in stones in the stream beds, where also iron-pyrites and probably chalcopyrite occur.[4] I have been unable to find these two latter minerals *in situ*, but they perhaps encouraged the gold mine interlude. Furthermore the galena very probably gave rise to trials about the same period, described below. This was a time of high silver prices when galena, being generally argentiferous, was regarded more as an ore of silver than of lead.

In 1676 John Claypole sought a lease to develop rich silver mines on two sites at Taynton Magna discovered by Abraham Shipton in 1673 when seeking coal.[5] Shipton appealed to the King but in 1677 all interests were assigned to Sir George Walker. Within two years he surrendered the lease, which passed to others. However, in 1685 Shipton was still trying to acquire the site and there is little doubt that active work was done. It seems very plausible to suppose that this abandoned spot with its overgrown 'quarry' and trial shaft is a memorial to these long-forgotten endeavours.

A—HEAD OF FRAME. B—FRAME. C—HOLES. D—EDGE-BOARDS. E—STOOLS
F—SCRUBBER. G—TROUGH. H—LAUNDER. I—BOWL.

This woodcut from Agricola's *De Re Metallica*, 1556, shows the kind of washing plant which probably served to extract gold in the Taynton trials three hundred years ago.

4 The Mines of Penyard Park

West of Newent, near Ross, is a prominent hill of Old Red Sandstone, which the Talbot family owned for centuries. It is skirted by 600 acres of woodland concealing many circular platforms where charcoal-burning took place to fuel iron-smelting furnaces from ancient times onwards.[1] Geologically, the ground is unpromising for mineral wealth, but nonetheless it discloses a number of very interesting trials of which practically nothing is recorded.

The woods also conceal extensive cliffs of conglomerate, so full of pebbles as fully to justify the old name, *Great Plum-pudding Stone*. These beds have attracted trials for gold only a mile or two away in the Forest of Dean; this probably explains an excavation in the cliffs behind Frogmore, where a level extends 30ft on a slight down-grade, and in conglomerate all the way. Its width is about 6ft and the present height 3ft, the floor being deep in earth. Probing near the entrance has proved an apparent void beneath the floor, which perhaps rests upon timbers. The general impression is one of great antiquity, and bearing in mind the proximity of Ariconium, we may suspect a Roman origin. The site deserves a proper investigation. Very probably further similar features will be found, although at present perhaps obscured by soil and undergrowth and fallen debris.

The other workings to be described are more recent, though their objects are somewhat obscure, and no doubt less ambitious. Beyond Lawns Farm are the

Driven into a rocky and inaccessible outcrop in Penyard Park, this level is probably an ancient trial for gold.

ruins of Penyard Castle, of which little remains other than the romantic shell of a woodward's cottage with stone mullioned windows and a fine chimney. If not once part of the castle, it was clearly built from its stones.

From this area a scrubby and rather impenetrable hillside descends with signs of activity from top to bottom. About 50yds down is an old shaft nearly filled with rubbish, and lower down are distinct hollows. Still lower, a Forestry road contours the hill, and close to where it has cut into the side was an adit or level, used some years ago for purposes of ritual. It has since been filled in or collapsed, but a hammer-shaped depression ascending the bank from the road no doubt marks the spot. According to an inspection in 1975,[2] it was '60ft long, about 5ft high and 5ft wide and dug by pick. At 4ft intervals it had had recessed wooden props on each side, set into walls which appeared to be Old Red Sandstone. The material the miners had sought was a grey soft glittering clay. This shone on our hands like silver paint. . . . The substance was a soft layer of rock, very friable, full of mica. . . . There were also 17 bats sleeping at the far end.' Nearly opposite, across the road and a little below, is an open vertical shaft, enclosed by a broken-down fence and covered by wire netting - dangerous, but of great interest to industrial archaeologists. It measures about 6ft × 8ft, and being oblong rather than round is probably not earlier than mid-19th century. The report of a descent in 1975 revealed bore-holes for blasting and 'every few feet down and across opposite walls were square holes that once held timbers.' The depth was about 70ft, the bottom being blocked by a layer of corrugated iron.[2] How much further the shaft descended is not known.

There is a considerable dump from the shaft, and adjacent to the west is a long gully terminating in another dump. It appears from the 1975 inspection that this trench is a fallen-in adit, and there may be other adits close by. Near the bottom of the bank below the shaft is a further dump, but its origins are uncertain.

From this area a valley runs down to the south, and a cart-track can be discerned descending parallel with a stream. Lower down and between the two is a distinct trench, short and deep (about 5ft), of the kind dug for prospecting purposes. These workings may not be all of a period, but it will be convenient to lump them together under the name Penyard Castle Mine.

The final workings are equally enigmatic. These may be termed the West Penyard Mine, and are again in woodland, just on the south side of the footpath across the hill, and about ½ mile west of Lawns Farm. A level has been driven obliquely up the hillside scarcely below the surface, so that over the years much has collapsed, leaving long gullies to mark its course. Nonetheless it can still be entered in places. Whatever was achieved might have been done easier by quarrying, so the method of working can only be described as very peculiar, regardless of the substance sought.

The level is driven in horizontal strata along a pronounced bed of tough chocolate-coloured clay or marl, above which is an ochreous sandstone with mica. Below, is a very crumbly and much more micaceous sandstone. Indeed, both this and the previous workings are known locally as mica or silver mines, however improbable that may be.

According to an informant who remembers it working as a boy,[3] West Penyard was privately financed by a Mr. Lennon or Leonard of Kilburn and two associates, having, it appears, been introduced to the spot by Forest miners about 1920 in the search for silver-sand, or sharp-sand. A tramline was laid down in the usual fashion, and the level went steeply down, 'like the roof of a house'. Bellows were used to supply air. However, as there is no sign of a dipple or inclined shaft, other unidentified workings may be indicated.

In short, these mysterious ventures in Penyard Park deserve a closer examination both on the ground and amongst the archives. Meanwhile, it is to be hoped that the physical testimony is accorded a proper respect, forming as it does, a major portion of Herefordshire's scanty contribution to mining in the border counties.

West Penyard Mine. Above; an entrance into the partially collapsed level.
Below, inside the level, showing the band of clay.

Notes and References

Recurring references are indicated as follows:

BGS	British Geological Survey.
CG	*Colliery Guardian.*
GJ	*Gloucester Journal.*
GRO	Gloucester Records Office.
GSIAJ	Journal of the Gloucestershire Society for Industrial Archaeology.
HRO	Hereford Records Office.
MJ	*Mining Journal.*
MGS	*Memoirs of the Geological Survey.*
Murchison Mss.	Manuscripts held by the Royal Geological Society of London.
Phillips	John Phillips, 1848, 'The Malvern Hills' *Memoirs of the Geological Survey, Vol 2, Part 1.*
PRO	Public Record Office, Kew.
QJGS	*Quarterly Journal of the Geological Society.*
Schubert	H. R. Schubert, 1957 *History of the British Iron and Steel Industry.*
TRGS	*Transactions of the Royal Geological Society.*

Introduction

1 National Library of Wales, Mynde Park 3082. **GRO**, D412 T1.
2 Coneybeare & Phillips, 1822 *Outlines of the Geology of England and Wales*, 428.
3 Coal Commission, 1871 Vol 2, 497.
4 Smith & Burgess, 1984 'The Permo-Triassic Rocks of the Worcestershire Basin'.
5 Cook & Thirlaway, 1955 'Measurements of Gravity in the Welsh Borders' **QJGS Vol 101.**
6 David Mushet, *Philosophical Magazine* Vol 40, 54-55.
7 **BGS** 1986. Dymock Forest Sheet Description (50 62NE). I hope to include details of the Mamble Coalfield in a forthcoming book on the Leominster Canal.
8 David Bick, 1979 *The Hereford & Gloucester Canal.*
9 Report of the Committee for the Hereford & Gloucester Canal, 1790.
10 **GJ** 19 July 1790.
11 *Hereford Journal*, 26 Jan 1791.
12 E. M. S. Paine 1861, *The Two James's and The Two Stephensons.* (Reprinted 1961).
13 David Bick, 1987, *The Gloucester & Cheltenham Tramroad.* (Oakwood Press).
14 David Mushet, *Phil. Mag.* Vol 40, 54-55.
15 **CG**, 11 July 1879.
16 J. Maclauchlan, 1838 **TRGS**, Series 2, Vol 5, 203, 4.
17 J. Douglas, 1912 *Historical Notes on Newent and District, 13.*
18 Information from Mr E. S. Morris of Ruardean.

Boulsdon

1 Brian Smith, 1976 'The Origin of Newent Coalmining' **GSIAJ**, 5-6.
2 **GRO**, D1803/6 map. See also Newent Tithe map and Apportionment.
3 Report of the Committee for the Hereford & Gloucester Canal, 1790.
4 Ralph Bigland, 1792 *Collections relative to the County of Gloucester.*

5 **GJ**, 6 Sept 1790, 4 April 1799.
6 **GRO**, D1810/2.
7 **GRO**, D1810/1.
8 Miles Hartland gave evidence to the Commissioners 'Report on Woods & Forests', 1792, wherein much on Forest collieries may be found. He died in 1796 and his tombstone is near the porch of St. Mary's Church, Newent. The Morse family tomb lies to the NE.
9 **GJ**, 6 Sept 1790.
10 G Turner, 1794 *General View of the Agriculture of Gloucester*, 55.
11 **Murchison Mss**.
12 David Mushet, *Phil. Mag.* Vol 40, 54.
13 T. Rudge, 1803 *History of the County of Gloucester*, Vol 2, 37.
14 T. Rudge, 1807 *General View of the Agriculture of the County of Gloucester*, 21.
15 *Ibid*, 338.
16 **GRO**, D245, 1-25
17 David Bick, 1987 *The Gloucester & Cheltenham Tramroad.* William Capel, died 1818, was an associate of Robert Hughes, see **GRO**, D245.
18 **Murchison Mss.**
19 **Murchison Mss.** refer to a shaft sunk 250yds deep at Boulsdon.
20 De la Beche's published account refers to the trial being 600ft to the south, but this seems an error (**MGS**, 1846, Vol 1, 214).
21 **CG**, 11 July 1879.
22 Maclauchlan refers to 'numerous' pits at Boulsdon.
23 According to John Phillips, Smith visited Boulsdon in 1805 but from his correspondence with Walter Honeywood Yate, August 1802 is more probable.

Kilcot

1 There is evidence that the survey for Price's map extended over many years, even into the 18th Century.
2 John Smith, *Men and Armour for Gloucestershire in 1608*, 63.
3 **GRO**, D1810/2.
4 **GRO**, D45 1/19.
5 **Phillips**, 105. Phillips was a nephew of William Smith, the geologist.
6 **PRO**, Hereford & Gloucester Canal Records.
7 G. Turner, 1794.
8 An old lady of Gorsley told me that a tunnel from the kilns led directly to the Oxenhall pits - no doubt a legend correct as to the source of coal, if not the route. For a description of the present-day remains (a fifth kiln was recently discovered), see David Bick, 1984 'Limekilns in North-West Gloucestershire'. **GSIAJ**, 2-12.
9 The capacity of a long-boat was 35 tons.
10 Miles Hartland's evidence to the Commissioners of Woods & Forests, 1792.
11 **GJ**, 5 Dec 1808. Perkins sold Wyatt's Farm at the same time.
12 **Murchison Mss**.
13 **Phillips**, 106.
14 De la Beche **MGS** 1846, Vol 1, 215.
15 **GRO**, D1882 contains many deeds and documents relating to Hill House, but I can find no reference to coal.
16 **MJ**, 7 June 1873.

17 Crooke's Mill, demolished early this century.
18 J. Douglas, 1912 *Historical Notes on Newent and District.*
19 *Cotteswold Naturalists Field Club* 1887, Vol 6, 132.
20 Information from the late Mr Baldwin of Pool Hill.
21 **GRO**, D 412 'Newent Notes'.
22 **BGS** Report, 1983 Vol 16, No 11. Murchison supposed the Boulsdon and Hill House main seams were the same, but the evidence tends to refute this.
23 H. Maclauchlan, 1838 **TRGS**, Series 2, Vol 5,
24 When Hill House Farm came up for sale in 1808 it included a tenement occupied by John George - probably a relative. South of Lower House is a red-brick dwelling where in 1842 stood a cottage occupied by Joseph George, thus confirming a tradition that the manager of the collieries lived here.
25 Information from Mr R. C. Goulding of Kilcot.

Oxenhall

1 **CG**, 11 July 1879. Whether much credence should be accorded these allegations may be doubted.
2 For details see David Bick, 1980, 'Remnants of Newent Furnace', **GSIAJ**, 29-37, also 'Remnants of Newent Ironworks', HMS 1982 AGM, 17-23. Regarding the steelworks see **GRO**, D2184.
3 For details see David Bick, 'Cast-iron Firebacks in West Gloucestershire' *Period Homes*, Sept 1985.
4 **GRO**. Will of R. F. Onslow.
5 **HRO**.
6 J. Phillips, 1805 *Inland Navigations.*
7 **MJ**, 7 June 1873.
8 G. Turner, *General View of the Agriculture of Gloucester*, 55.
9 For details of the Moggridge and Perkins families see H. S. Torrens, 'Geological communication in the Bath area in the last half of the eighteenth century' Chap. 10, 215-247. *Images of the Earth*, British Society for the History of Science, 1979.
10 P. G. Rattenbury, 'The Penllwyn Tramroad', *R & CHS Journal* Vol 27, No 7 189-197.
11 B. C. James, 'J. H. Moggridge and the founding of Blackwood', *Presenting Monmouthshire* No 25, 26-29.
12 According to Michael Handford such papers still exist.
13 **PRO**, H & G Canal Joint Committee Minutes.
14 **MJ**, 7 June, 1873.
15 For a summary, see David Bick, 1979 'Records of the Newent Coalfield' **GSIAJ**, 1-8.
16 **Phillips**, 106.
17 **GJ**, 30 June, 14 July 1877.
18 **CG**, 11 July 1879.
19 This would be one of the Newent Colliery companies.
20 L. Richardson, 1942 'Notes on the Geology of the Newent District' *Trans Woolhope Nats. F.C.* 47-52.
21 W. S. Symonds, 1857 *Stones of the Valley*, 66.
22 **Phillips**, 108.
23 Parliamentary Plan of Newent Railway.
24 **GRO**, Oxenhall Tythe map.
25 **CG**, 11 July 1879.

26 *Cotteswold Naturalists Field Club*, Vol 6, 130-3.
27 Gethyn-Jones, 1951 *Dymock Down The Ages*, 179.
28 Information from the late Mr Huff, son-in-law of Charles Jones.
29 R. I. Murchison, 1839 *The Silurian System*, 154.
30 **PRO**, H & G Minutes 6 Sept 1798.

Newent Colliery
1 Morris & Co, 1876 Trade Directory.
2 **PRO**, B T 31, 2234/10599, X/K 1726, upon which much of this chapter is based.
3 Boring gear was taken from Boulsdon to Oxenhall about this time. See **GJ**, 24 Feb 1953.
4 **GJ**, 12 July 1878.
5 **CG**, 18 July 1879.
6 Information from the late Mr Baldwin.
7 Other trustees included the Rev. Robert Burroughes of Pencombe and Major W. C. Hill of Powick, sons-in-law of R. F. Onslow.
8 **GJ**, 20 Sept 1879.

Dymock
1 The road here was once so narrow that vehicles could not pass.
2 **Phillips**, 107.
3 David Bick, 1980 *Field Notes of Old Mines*, Vol 4, 38-41. (Manuscript).
4 Papers held in Gaveller's Office, Coleford.
5 I am indebted to Mrs V. Marshall for reminiscences of her late husband.
6 R. I. Murchison, 1893 *The Silurian System*, 154.
7 **GRO**, D185 N10.
8 Document in possession of David Nunn, Little Woodend.
9 Smith Mss, Oxford.
10 **GJ**, 2 May 1803, 14 April 1805. See also R. Newman, 1984 *Coal Mining at Dymock*, **GSIAJ.**

Trials for Coal
1 **Phillips**, 54.
2 S. Rudder, 17779 *A New History of Gloucestershire* Appendix 23.
3 **GRO**, D 1938.
4 **GRO**, Tithe map.
5 **MJ**, 5 Oct 1867.
6 For a history of coal prospecting see H. S. Torrens, *Energie in der Geschichte*, papers presented to the 11th International History of Technology Symposium. Dusseldorf 1984, 88-95.

Iron Mining and Ironworks
1 A. Yarranton, 1677 *England's Improvement on Land and Sea*, Part 1.
2 Information derived from Tithe maps.
3 See especially **Schubert** and Cyril Hart, 1971 *The Industrial History of Dean.*
4 David Mushet, 1840 *Papers on Iron and Steel*, 389.
5 But see John van Laun, 1979 '17th Century Ironmaking in South-west Herefordshire', *Historical Metallurgy* Vol 13 No 2, 55-68.

6 B. L. C. Johnson, 1953 'New Light on the Iron Industry of the Forest of Dean', *B & GAS* Vol 72, 136.
7 These papers have inspired a number of important studies, and promise much more.
8 Johnson, 'New Light . . .', 135.
9 R. A. Stiles, 1971 'Elmbridge Furnace, Oxenhall' *Gloucestershire Historical Studies*, 2-11. This is a useful paper but its calculations on yields are dubious.
10 The iron content of cinders varied greatly from place to place; See Hart, 69.
11 **Schubert**, 244.
12 Johnson 'New Light . . .'
13 **Schubert**, 244.
14 For other locations see N. P. Bridgewater, 1968 'Iron-making and Working Sites in and around the Forest of Dean' *H.M. Group Bulletin* Vol 2 No 1, 27-32.
15 For the local geology see J. D. Lawson, 'The Geology of the May Hill Inlier' **QJGS** Vol CX1 Part 1, 85-100. See also the relevant **BSJ** notes, in course of publication.
16 **GRO** Foley estate map.
17 The diary is deposited in Hereford Records Office.
18 N. P. Bridgewater, 1965 'Roman-British Iron Working near Ariconium', Woolhope NFC, Vol 38(2), 124.
19 **HRO** 974 LC.
20 For some interesting observations, see Brian Cave, 1982, *Weston & Lea* (Forest Bookshop).
21 **Shubert**, 385-6.
22 Trans Woolhope Nats. Field Club, 1936, 70.
23 **HRO**, 974 LC.
24 Discovered by Mr Howard Ellis of Ross.
25 The possibility that this point may have constituted the exit of the tunnel from a water-wheel pit should be borne in mind.
26 **HRO** F/vi/DDF/4.
27 I have not thoroughly explored between Pinford and Fording farms. Interestingly enough, just SW of Shutton near the old turnpike road to Ross, is a high-level leat and header-pond as if for hushing, and revealing hallmarks of antiquity.

Trials for Silver and Gold

1 This chapter is based on David Bick, 1986 'The Mines and Minerals of Newent', *Journal of the Russell Society*, Vol 1 No 4, 114-118.
2 Abbot, 1833, *Essay on Metallic Works*, 203.
3 Cyril Hart, 1971 The Industrial History of Dean, 316-321.
4 The occurence was revealed about 1955 during a search for the 'gold mine'.
5 W. Rees, 1968 *Industry before the Industrial Revolution*, Vol 2, 470.

The Mines of Penyard Park

1 W. H. Cooke, 1882, *County of Hereford*, 220. See also Trans. Woolhope NFC, 1927, 134; 1930, 80.
2 Rollo Gillespie, 1975, 'Ross Silver Mines', *Royal Forest of Dean Caving Club*, 13-17.
3 Information from Mr Wally Hall of Ross. The gentleman of Kilburn was married to his aunt Jane. Mr Hall's father and brother worked in the mine and afterwards opened a sand quarry at Knacker's Hole Grove, from which an inclined plane with a passing loop conveyed trams to a landing-stage at the roadside west of Bill Mills. This operated for several years.

Sources and Acknowledgements

Time has dealt severely with records of Newent mining, and details of primary material occupy but little space. These comprise the canal company minutes, the Foley Papers at Hereford, plus a few items in Gloucester Records Office and the Gaveller's Office at Coleford.

Trade periodicals, maps and geological and topographical works together with local papers have proved the most useful source. The Public Records Office holds valuable evidence on Newent Colliery but in general the story of individual pits is almost a closed book.

Compiling an account of this kind involves a continual search for pieces of a complex historical jig-saw of which nearly all are irrevocably lost, with in the end the daunting task of attempting to resurrect a truthful and balanced picture. In these endeavours so many have helped me that it is scarcely possible to name them all. In particular I have to thank my old friend George Hall for references to the *Mining Journal*, Jeremy Wilkinson for transcripts from the Public Records Office and Elliot Evers for photographs and details of family connections with Newent Colliery.

For drawing attention to the culverts and tunnels in the vicinity of Burton Court, and for permission to examine them, I am much indebted to Mr E. R. Vines and Mr C. F. Huntley.

I am also grateful for a variety of assistance to Roger Aston, Don Barton, Laurence and Roger Bailey, A. Bevan, Dr Ivor Brown, Mike Byers, Brian Cave, Amina Chatwin, Trevor Chesters, C. L. Cleal, John Cornwell, Lyndon Cummins, A. J. Davies, A. H. Done, R. H. Elgie, Howard Ellis, Eric Evans, Paul Eward, C. Farnham, Rollo Gillespie, Harry Garratt, K. Goulding, R. Goulding, J. J. Haden, Charles Hadfield, Wally Hall, F. C. Herrick, Commander M. R. D. Hooke, Albert Howell, G. Hurst, Dr R. J. King, J. T. Lowe, James Lowe, Mrs V. Marshall, Jack Morrell, E. S. Morris, Reg Nicholls, David Nunn, Mrs Enid Parker, Dood Pearce, Donald Pennington, Mrs F. Penney, Mrs Z. Sayell, Dr J. A. Secord, Dr T. Sharp, Alan Simpson, Brian Smith, Peter Smith, Ian Standing, Robin Thewlis, Dr Hugh Torrens, Mrs Gwen Tutt, Jenny Whiskerd, Bernard Worssam and Ray Wright. For the drawing of Newent Colliery on the cover I am indebted to Michael Blackmore of Abergavenny.

Research has involved many contacts with public institutions and I should like especially to thank the staffs of Gloucestershire Records Office and Gloucester Reference Library for their unfailing courtesy on every occasion. I must also acknowledge help from the staffs of Birmingham Public Library, the British Library and Museum, the Bodleian Library, Stourbridge Reference Library, Hereford & Worcester, Shropshire, Somerset and Staffordshire Record Offices, also the House of Lords Record Office. Thanks are also due to the staff of the Geological Society of London and the British Geological Survey. Finally a word of appreciation for my family without whose tolerance of frequent absences and many an evening's preoccupation, this little work would not have seen the light; also for Rose Maynard, for the tedious task of correcting the proofs.

Select Bibliography

1779 S. Rudder, *A New History of Gloucestershire.*

1791 R. Bigland,*County of Gloucester.*

1794 G. Turner, *General View of the Agriculture of Gloucester.*

1803 T. Rudge, *The History of the County of Gloucester.*

1807 T. Rudge, *Agriculture of the County of Gloucester.*

1837 J. Maclauchlan, 'Notes to accompany a geological map of the Forest of Dean Coalfield'. *Trans. Geol. Soc.,* Ser 2, Vol 5.

1839 R. I. Murchison, *The Silurian System.*

1846 H. De la Beche, 'On the Formation of Rocks in South Wales and South Western England'. *Mem. Geol. Survey* Vol 1.

1848 J. Phillips, 'The Malvern Hills . . .' *Mem. Geol. Survey* Vol 2, pt 1.

1857 W. S. Symonds, *Stones of the Valley.*

1910 N. Arber, 'Notes on a Collection of Fossil Plants from the Newent Coalfield'. *Geol. Mag.* New Series. Decade V. Vol VII.

1912 J. Douglas, *Historical Notes on Newent and District.*

1930 L. Richardson, *Wells and Springs of Gloucestershire.*

1942 L. Richardson, 'Notes on the Geology of the Newent District', Trans. Woolhope N.F. C

1944 C. E. Hart, 'Gold in Dean Forest'. *B. & G. A. S.*

1951 J. E. Gethyn-Jones, *Dymock Down The Ages.*

1953 B. L. C. Johnson, 'New Light on the Iron Industry of the Forest of Dean'. *Bristol & Gloucester Arch. Soc.* Vol 72, 129-143.

1955 J. D. Lawson, 'The Geology of the May Hill Inlier'. *QJGS,* Vol 111.

1971 D. E. Bick, 'The Newent Coalfield', *Glos. Historical Studies.*

1976 B. S. Smith, 'The Origin of Newent Coalmining', *Glos. Soc. Ind. Arch. Journal.*

1979 D. E. Bick, 'Records of the Newent Coalfield', *Glos. Soc. Ind. Arch. Journal.*

1979 D. E. Bick, *The Hereford & Gloucester Canal.*

1984 R. Newman, 'Coal Mining at Dymock', *Glos. Soc. Ind. Arch. Journal.*

1986 B. S. P. Moorlock & B. C. Worssam, 'Geological Notes for 1:10,000 Sheet SO 62 NE (Dymock Forest), *BGS.*

1986 D. E. Bick, 'The Mines and Minerals of Newent', *J. Russell Soc.* 1, 114-118

APPENDIX I

Coal Output and Reserves

Although further workings except on a very small scale can be ruled out for various reasons, nevertheless it is tempting to speculate on the reserves of the Newent Coalfield. Unfortunately, one part of the district is difficult or impossible to correlate with another, and the total thickness of the Coal Measures has never been ascertained. The boreholes which the British Geological Survey recently drilled 860ft through inclined strata at Kilcot neither began at the top nor reached the bottom of the measures. However, allowing for dip, the true thickness proved at least 690ft containing not less than 20ft of coal, or a ratio of 60:1.

The simplest method of tackling the question is to make a few basic assumptions, which can readily be modified as desired. The Coal Measures will be presumed to occupy two square miles (including areas covered by New Red sandstones), with a single seam 6ft thick, of specific gravity 1.25. Such a seam contains 1,520 tons per acre for every foot in thickness, from which the total comes to 11,670,000 tons. Of this, but a fraction has been extracted - a not unusual state of affairs in coalfields with difficult geological problems.

As an example we may take the Bristol and Somerset Coalfield, abandoned about 1972 after a long and chequered history. over a century ago Professor Prestwich estimated it still contained 3,238,789,000 tons of coal, workable or otherwise, not more than 3,000ft deep. Of this, scarcely 3% has since been mined.

As to the total output of the Newent Coalfield such a compilation can only be hazarded, but there are several useful clues.

Boulsdon

Presuming coal to the deep of the shaft behind behind Great Boulsdon was left untouched, about 20 acres must have been worked. For a 5ft seam and 40% extraction, the yield would be 61,000 tons. Since the pits were active, on and off, at least from 1608 till 1810, this figure is I think a safe one.

Footnote:

Whilst these pages are in the press, coal from the field opposite Boulsdon Croft has been examined by a competent authority, and pronounced as perfectly acceptable house-coal, rather low in calorific value, but with good ash and sulphur characteristics. This unexpected result raises a host of questions, too late for consideration here. The Boulsdon pond coal is reported as poor, but very different from the Lower House material.

Kilcot
Of output before the 1790s, nothing can be deduced.

1792-1800
Up to the end of 1795 the Hereford & Gloucester Canal records provide three independent sources - revenue from sales, payments for haulage, and royalties. Depending on assumed prices, by each method the quantity raised and sold comes to about 1,000 tons/year. If this rate continued we can assess the total at 8,000 tons.

However, Maclauchlan's evidence in referring to workings extending for 300yds suggests a higher figure. With a depth of 102ft to the 'Big Coal' this permits another estimate, assuming a dip of 18ins/yd and a 6ft seam. A 40% extraction would yield about 16,000 tons, ignoring produce from thinner seams. Hence, on average, we may presume 12,000 tons.

1839-46
Mining continued in the Perkes interlude to a much greater depth, judging by the shaft stated to be about 100yds deep. The logic of the above calculations indicates a further 40,000 tons, or say 20,000 tons to be safe.

Thus the above figures suggest 32,000 tons for Kilcot, without taking into account production before 1790 or in the 1870s.

Oxenhall and Dymock
For these parishes we can only guess. The pits, of which some were never much more than a name, included White House, Oxenhall, Newent and Court Hill Collieries, also sundry workings and trials as at Peter's Farm, Hillend, Castle Tump and beyond. Assume 8,000 tons altogether.

Total Output of Newent Coalfield
The above figures can be summarised as follows

Boulsdon	61,000 tons
Kilcot	32,000 tons
Oxenhall, Dymock	8,000 tons
	101,000 tons

Future Prospects
It was a firm belief amongst old Forest miners that until the Forest of Dean's boundaries became contracted, their rights included the Newent Coalfield. At all events, the possibilities of a little mining on a 'one man and a boy' basis should not be ruled out, if only to supply the landowner's hearth with fuel. A resurrection, if only on the smallest scale, would be of more than a little scientific and historic interest.

APPENDIX 2

Locations of Workings and Trials

1 The Newent Coalfield

Name	*Grid Ref.*	*Parish*	*Mineral*
Boulsdon*	710245	Newent	Coal
Wyatt's	704249	Newent	Coal
Lower House Colliery*	698262	Oxenhall/Newent	Coal
Hill House Colliery*	698263	Oxenhall	Coal
Colliery Meadow	697261	Oxenhall	Coal
Newent Colliery*	700267	Oxenhall	Coal
White House Colliery	699267	Oxenhall	Coal
Aston's Slope	699270	Oxenhall	Coal
Peter's Colliery*	703271	Oxenhall	Coal
Water-level?	704271	Oxenhall	Coal
Court Hill Colliery	703274	Oxenhall	Coal
Holders	705276	Oxenhall	Coal
Hillend Level*	706283	Oxenhall	Iron Ore
Hillend Pond	706284	Oxenhall	Coal
Castle Pond	711292	Dymock	Coal
Castle Tump	712293	Dymock	Coal
Coalpit Field	713297	Dymock	Coal
Pit Leasows	716300	Dymock	Coal

2 Workings outside the coalfield

Frogmore	629220	Weston	Gold?
Penyard Castle	619223	Ross	?
West Penyard	613228	Ross	Silver-sand?
Aston*	691235/7	Aston Ingham	Iron ore
Crocket's Hole	719232	Newent	?
Black House*	716225	Newent	Iron ore
Lead Mine	715220	Newent	Silver
Rock's Wood	717221	Newent/Taynton	Gold
Coal Hill	757356	Bromesberrow	

Notes

1 An asterisk denotes sites known to have sold coal or ore.
2 Most sites are on private property and at some, no physical evidence remains.

APPENDIX 3
Sections of Shafts and Borings

BOULSDON | KILCOT | OXENHALL

50' 100' 150' 200' 250' 300' 350' 400' 450'

SMITH 6·0 3·3
MURCHISON 1·5 0·8 2·5
DE LA BECHE 5·5

HILL HOUSE — MACLAUCHLAN 7·0 2·5 4·0
LOWER HOUSE — MURCHISON 7·0 2·0 1·5 1·2
BGS BORING 0·2 1·9 2·6 1·0 1·0 1·1 2·5 — 5 seams in a further 450'.

1791 BORING 2·5 1·1 2·7
WHITE HOUSE 7·5 HEMATITE 0·3 2·0 IRON-STONE 0·4 6·7 2·4
NEWENT COLLIERY 6·5 1·5

INDEX